Intelligent Agents for Mobile and Virtual Media

Springer
London
Berlin
Heidelberg
New York
Barcelona
Hong Kong
Milan
Paris
Singapore
Tokyo

Rae Earnshaw and John Vince (Eds)

Intelligent Agents for Mobile and Virtual Media

 Springer

Rae Earnshaw
Department of Electronic Imaging and Media Communications,
University of Bradford, Bradford, BD7 1DP

John Vince
Media School, Bournemouth University, Talbot Campus,
Fern Barrow, Poole, BH12 5BB

British Library Cataloguing in Publication Data
Intelligent agents for mobile and virtual media
    1.Virtual computer systems 2.Mobile computing 3.Intelligent
    agents (Computer software) 4.User interfaces (Computer
    systems)
    I.Earnshaw, R. A. (Rae A.) II.Vince, John, 1941-
    006.3
    ISBN-13: 978-1-4471-1175-7

Library of Congress Cataloging-in-Publication Data
A catalog record for this book is available from the Library of Congress

ISBN-13: 978-1-4471-1175-7   e-ISBN-13: 978-1-4471-0677-7
DOI: 10.1007/978-1-4471-0677-7

a member of BertelsmannSpringer Science+Business Media GmbH
http://www.springer.co.uk

Typesetting: Ian Kingston Editorial Services, Nottingham
Printed and bound by the Athenæum Press Ltd., Gateshead, Tyne and Wear
34/3830-543210 Printed on acid-free paper SPIN 10853756

Contents

List of Contributors

Giovanni Aloisio
Department of Innovative
Engineering
University of Lecce
Via per Monteroni
73100 Lecce
Italy
Email:
giovanni.aloisio@unile.it

Ruth Aylett
Centre for Virtual
Environments
University of Salford
Salford M5 4WT
UK
Email:
r.s.aylett@salford.ac.uk

Daniel Ballin
Radical Multimedia Laboratory
BTexaCT
Admin 2–5
BT Adastral Park
Ipswich IP5 3RE
UK
Email:
daniel.ballin@bt.com

Simon Bramble
The Forensic Science Service
London Laboratory
109 Lambeth Road
London SE1 7LP
UK
Email:
sbramble@hgmp.mrc.ac.uk

Nic Chilton
Department of Electronic
Imaging and Media
Communications
University of Bradford
Bradford BD7 1DP
UK
Email:
N.Chilton@bradford.ac.uk

Wen Hao Chuang
MIC – Multimedia Innovation
Centre
The Hong Kong Polytechnic
University
Hung Hom
Kowloon
Hong Kong
Email:
cswen@polyu.edu.hk

L. K. Comerford
Microcomputer Music Research
Unit
School of Computing and
Mathematics
University of Bradford
Bradford BD7 1DP
UK
Email:
l.k.comerford@scm.brad.ac.uk

P. J. Comerford
Microcomputer Music Research
Unit
School of Computing and
Mathematics
University of Bradford
Bradford BD7 1DP
UK
Email:
p.j.comerford@scm.brad.ac.uk

Jurgen Dabeedin
The Forensic Science Service
London Laboratory
109 Lambeth Road
London SE1 7LP
UK
Email:
jad03@fss.org.uk

Carlos Delgado
Centre for Virtual
Environments
University of Salford
Salford M5 4WT
UK
Email:
c.delgado@pgr.salford.ac.uk

Peter S. Excell
School of Informatics and
Telecommunications Research
Centre
University of Bradford
Bradford BD7 1DP
UK
Email:
P.S.Excell@Bradford.ac.uk

Christos Georgousopoulos
Department of Computer
Science
Cardiff University
POBox 916
Cardiff CF24 3XF
UK
Email:
geolos@cs.cf.ac.uk

M. Gerhard
School of Computing
Leeds Metropolitan University
Leeds LS1 3HE
UK
Email:
m.gerhard@lmu.ac.uk

Mark Goodall
Department of Electronic
Imaging and Media
Communications
University of Bradford
Bradford BD7 1DP
UK
Email:
markg@bilk.ac.uk

D. J. Hobbs
Department of Electronic
Imaging and Media
Communications
University of Bradford
Bradford BD7 1DP
UK
Email:
d.hobbs@bradford.ac.uk

Patrick Ingham
School of Computing
Leeds Metropolitan University
Leeds LS1 3HE
UK
Email:
p.ingham@lmu.ac.uk

Mikael Jern
ITN
Linkoping University
Sweden and Advanced Visual
Systems
Denmark
Email:
mikael@avs.dk
mikje@itn.liu.se

R. S. Kalawsky
Advanced VR Research Centre
Department of Computer
Science
Loughborough University
Loughborough LE11 3TU
UK
Email:
r.s.kalawsky@lboro.ac.uk

D. J. Moore
School of Computing
Leeds Metropolitan University
Leeds LS1 3HE
UK
Email:
D.Moore@lmu.ac.uk

Ian Palmer
Department of Electronic
Imaging and Media
Communications
University of Bradford
Bradford BD7 1DP
UK
Email:
i.j.palmer@bradford.ac.uk

Jon Pettigrew
Creativity and Cognition
Research Studios
LUTCHI Research Centre
Department of Computer
Science
Loughborough University
Loughborough LE11 3TU
UK
Email:
j.s.pettigrew@lboro.ac.uk

Omer F. Rana
Department of Computer
Science
Cardiff University
PO Box 916
Cardiff CF24 3XF
UK
Email:
o.f.rana@cs.cf.ac.uk

Donna Robey
The Forensic Science Service
London Laboratory
109 Lambeth Road
London SE1 7LP
UK
Email:
dro03@fss.org.uk

Karen D. Scott
National Museum of
Photography, Film & Television
Bradford
West Yorkshire BD1 1NQ
UK
Email:
K.Scott@nmsi.ac.uk

Wen Tang
School of Computing and
Mathematics
University of Teesside
Middlesbrough
TS1 3BA
UK
Email:
W.Tang@tees.ac.uk

Roy Williams
Center for Advanced
Computing Research
California Institute of
Technology
Pasadena
California
USA
Email:
roy@caltech.edu

David W. Walker
Computational Sciences
Section
Oak Ridge National Laboratory
PO Box 2008
Oak Ridge
TN 37831-6367
USA
Email:
walkerdw@ornl.gov

Tao Ruan Wan
Department of Electronic
Imaging and Media
Communications
School of Informatics
University of Bradford
Bradford BD7 1DP
UK
Email:
T.Wan@bradford.ac.uk

Anne M. White
Department of Modern
Languages
University of Bradford
Bradford
BD7 1DP
Email:
a.m.white@brad.ac.uk

Yanyan Yang
Department of Computer
Science
Cardiff University
PO Box 916
Cardiff CF24 3XF
UK
Email:
Linda@cs.cf.ac.uk

Preface

The marriage of communication and computer technology is responsible for the current phenomenon of digital convergence, where digital systems are destined to become ubiquitous features of our daily lives. Designers of such systems are fully aware of the technical complexity required to design them, especially at the user interface. Not only must these systems be useful, they must be commercially viable; otherwise their life will be rather short.

Today, the Internet and the World Wide Web are finally making an impact on corporate and private behaviours, where email, virtual environments, digital video electronic shopping etc. are becoming everyday activities. But as these systems grow larger and more complex, we must minimise the technological competence expected from the system user – consequently they must become much more intelligent.

Modern PCs are still far from intelligent or user-friendly and can only be used by people with sufficient tenacity to cope with the frustration caused by badly designed and unreliable software. And the term "intelligence" is far from anyone's mind when using such software. Nevertheless, in spite of such criticisms, no one can deny the latent power waiting to be released from computer-based systems.

In order to simplify the *human-to-computer* interface, systems are needed that mimic the *human-to-human* interface, with which we are so familiar. Hopefully *intelligent software agents* will play a significant role in the design of such interfaces, and in this book we explore how they could influence media-based systems.

In Chapter 1, E. H. Mamdani writes about *Convergence, Software Agents and Information Ecosystems*. Professor Mamdani clarifies the term "convergence" and explores the interface issues associated with *human-to-machine* and *machine-to-machine*. It is the latter mode of communication that still has to be defined and will eventually open the way to silicon life, as hinted at towards the end of the chapter.

In Chapter 2, Roy Kalawsky reports on work being undertaken at the Advanced VR Research Centre at Loughborough University in the

UK. Professor Kalawsky has summarised his findings in *Interface Issues for Wearable and Mobile Computer Users*, where he explores the wide range of interface problems associated with small mobile technologies.

One of the main uses of computer-based systems involves the retrieval of data, whether it is the location of services and products or statistical information about numerical data sets. Visualising such data has proved to be a real success for computer graphics, and in Chapter 3 Professor Mikael Jern surveys the latest navigational tools available to online users in *Visual Data Navigators "Collaboratories" True Interactive Visualisation for the Web*.

In Chapter 4, Professor Peter Excell's *The Future of Convergent Computer and Telecommunications Technology* reviews, and raises questions about the way in which personal computer and mobile communications technologies will be used in the near/medium-term future. Professor Excell reviews the evolutionary paths of the relevant technologies and argues that they are converging rapidly.

For many years, computer animators have investigated ways of making their characters behave more realistically, and this goal for extra realism has also been pursued in real-time virtual environments. In collaboration with the Centre for Virtual Environments at the University of Salford, British Telecom's Radical Multimedia Laboratory has been researching into intelligent virtual agents. The result of this research is described by Daniel Ballin, Professor Ruth Aylett and Carlos Delgado in Chapter 5, entitled *Towards Autonomous Characters*.

Chapter 6 describes a practical example of where an agent-based architecture provides tangible benefits over more traditional software engineering techniques. In *Towards an XML and Agent-Based Framework for the Distributed Management and Analysis of Multi-spectral Data*, Omer Rana and colleagues report how their agent-based system offers the advantages of modularity, scalability, decentralisation and extensibility.

Pursuing a similar theme to that of Chapter 5, Tao Wan and Wen Tang in *Learning by Experience – Autonomous Virtual Character Behavioural Animation* present a novel goal-oriented approach for complex virtual character behavioural simulation. A central feature of their work is the use of genetic algorithms to optimise the parameters used for animating their characters.

The subject of Chapter 8 is *The Development of an Intelligent Virtual Environment for Training*. It is the result of an exciting collaborative project between the Forensic Science Service in London and the Department of Electronic Imaging and Media Communications at the University of Bradford. Ian Palmer and colleagues explore the potential use of real-time VEs as an additional tool in crime scene investigation.

Anticipating that avatars will play an important role in collaborative VEs, Dr David Hobbs from the University of Bradford has
teamed up with Dr Moore and Michael Gerhard from the Leeds
Metropolitan University to evaluate how to measure presence in a
VE. *An Experimental Study of the Effect of Presence in Collaborative
Virtual Environments* (Chapter 9) reports on their research with the
conclusion that animated realistic avatars do create more presence
than basic shape avatars.

It is hoped that distance-based learning and Web-based instruction
will play a significant role in future international teaching programs.
At the Multimedia Innovation Centre at the Hong Kong Polytechnic
University, Wen Hao Chuang has been researching into how such
strategies can be used for teaching music. The results of this research
are reported in Chapter 10: *Formative Research on the Refinement of
Web-Based Instructional Design and Development Guidance Systems
for Teaching Music Fundamentals at the Pre-College Level.*

The BBC's series *Walking with Dinosaurs* was a watershed in television broadcasting and paved the way for computer animation to
become a major aspect of future television programs. The authors
of Chapter 11, Karen Scott and Anne White, analyse in *Unnatural
History? Deconstructing Walking with Dinosaurs* how the TV series
stands up to critical analysis.

Mark Goodall in Chapter 12, *Trashing the Net: Subcultural Practice
Online*, critically examines uses of the WWW by fans of cult movies.
Examples are presented to illustrate the question of whether,
through remediation processes, such practices tell us anything new
about forms of contemporary communication and consumption.

In Chapter 13, *Another Time, Another Place: the Use of Technology to
Make the Cultural Heritage of the Organ Accessible*, L. K. and P. J.
Comerford discuss different ways in which digital technology can
be used by music students to illustrate different organ cultures. This
includes the use of sound synthesis technology designed specifically to take into account these requirements.

The title of Chapter 14, *Synthetic Vision for Road Traffic Simulation
in A Virtual Environment*, is the result of a collaborative project
between Wen Tang at the University of Teesside and Tao Ruan Wan
at the University of Bradford. The paper describes a method of
using synthetic vision for simulating virtual traffic and evaluating
driver behaviour.

Finally, in Chapter 15 Jon Pettigrew from the LUTCHI Research
Centre at Loughborough University explores the usage and attitudes of children using computers for creative collaboration. The
Eternity project is a *"gedanken"* art project (that is, an art "thought
experiment") and invites children aged 10–12 to answer the question "How can we create a piece of music to last for ever?".

The central theme for these chapters is the role of intelligent software agents and similar programming strategies that can be used to simplify the *human-to computer* interface. And when you have read these chapters you will discover that considerable advances have already been made. But we are only at the beginning of a long journey towards a day when computers will be truly described as an intelligent aid to our personal and business lives.

Rae Earnshaw
John Vince

1

Convergence, Software Agents and Information Ecosystems

E. H. Mamdani

1.1 Introduction

This chapter is about convergence and its probable future impact upon our society. The effects of convergence can already be seen in the world's economy and society. Many of the current events are transient phenomena that will not last, while others give a clear hint of the trends and suggest which way we are likely to go in the future. Unfortunately, it is not easy to tell the lasting from the transient while we are in the middle of the ongoing developments. All we can say is that something significant and important for the future is taking place as a result of advances in digital technologies and communications networks. Our aim in this chapter is to try to step back from current developments in the hope that we can see some of the underlying trends.

Below, we first describe how convergence has come about and then discuss one key further development of it. This is that independent, intelligent, decision-making computational entities – often called software agents – may begin to make use of the digitised communications infrastructure. Section 1.4 then discusses briefly what experience is being gained by those who have exploited some of the possibilities of convergence. Finally, we try to look further into the future, which we call here the digital information ecosystem.

1.2 Convergence

The term "convergence" refers to the strong coupling that has begun to emerge between what until recently were quite separate lines of communications-related businesses: telephony; audio-visual broadcast networks; and computer networks. Digitisation of content and the ability to transport this content is at

1

the heart of convergence. It is worth briefly tracking the progress towards convergence that has been taking place over the past couple of decades.

The core of the telephone network and the associated switching infrastructure was the first to go digital. The telephone network had always provided the required connectivity for connecting terminals to mainframe computers. With personal computers, terminals became less dumb and we went from 1000 people sharing one computer to today's situation, in the developed world, of several computers per human user. So when personal computers arrived, the long-distance transport of computer data was implemented by leasing the already digitised lines from the telephone companies. Locally, computers were connected by purpose-installed wiring and using a number of protocols, but chiefly the Ethernet protocol. Telephone companies were happy to get this new business of leased digital lines and left it to the computer industry to route their own traffic at the edges of the phone network. The creation of that computer network required the generation of a whole suite of what came to be known as the Internet Protocols.

Today, of course, even mobile operators use the network resources set up by the incumbent phone companies. This situation, in which most of the traffic carried originates and even terminates outside their networks, is apparently not entirely to the liking of incumbent phone companies. Regulators, who are mainly concerned with current issues talk of mobile/fixed convergence and computer/telephony convergence. Our concern here is not with present-day issues of convergence but with examining trends within what is clearly a disruptive technology. Disregarding the booms and busts within the TMT (Technology, Media and Telecom) sector of the economy, it should be noted that digitisation has introduced new levels of uncertainty in the economy that will take a while to settle down. Furthermore, we must not forget that the audio-visual industry has also been busy creating its patches of the access network, mainly for carrying TV broadcast channels; but that network is also offered for carrying telephone and data traffic. What the future impact of the digitisation of the audio-visual broadcast industry will be on converging with the telephone and computer parts of the market still remains to be seen.

In fact, digital technology is potentially even more disruptive than the above discussion suggests, because the technology offers more than just the digitisation and transport of that content. It offers the ability to store vast quantities of that content and to process it as well. While digital broadcast content is mainly pushed out to the users, a user mostly pulls the computer-based digital content when it is needed. The large server farms that have been created have extraordinary traffic patterns. Gigabytes of information radiate out of these server farms, but comparatively little information goes into them. Asymmetric traffic patterns exist within telephone networks also, but they are nothing like those that occur in data networks. Indeed, traffic patterns within the whole area of communications technologies and the way they evolve give a very clear indication of what the rhythm of a given community is and how that rhythm is evolving. At present, the networks are not being used efficiently because most of the information being sent out is exactly the same, but individually repeated for each user that requests it and staggered in time in response to the requests received. Technical solutions are being devised so that all requests can be gathered together and information pushed out simultaneously to each of the requesting user.

1.3 Software Agents

With *human-to-human* communication there are now a number of alternatives available to replace the telephone- or letter-based communication of only a few years ago. The business impact of such a substitution is that new players have provided the alternative forms of communication services (email, mobile phones etc.), which have consequently taken much of the business away from the incumbent phone companies and the postal services. The past year has marked the end of the phone companies' cash cow – fixed line telephone traffic actually declined in 2000.

As we go into the future, we need to take into account two new forms of entity that exist and which will participate in the majority of the communications taking place. The *human-to-human* communication that we all think of when we talk of communications is likely to become a small percentage of the total. We all know that the Web has been remarkably successful in providing easy electronic access to largely textual information. This partial substitute for paper-based archives can be expected to get more sophisticated in the future. *Human-to-archive* communication already accounts for a large part, if not the majority, of network traffic.

Moreover, we can already see that an increasing amount of Web access is not just access to information, but takes the form of an automatic transaction with some organisation's computers. This is a form of *human-to-machine* communication that can be expected to grow. All e-commerce is essentially *human-to-machine* communication.

An obvious next step is to see machines communicating with other machines without any intervention from humans. There are many problems that need to be solved to allow this form of communication to work to society's advantage. These problems are at the heart of research taking place in what is termed software agents. In essence, the application of software agents is really another form of automation that we have all got used to over the past 50 years. Software agents are designed to "meet" one another on the open Internet and carry out transactions acting on their owner's behalf. Quite a large amount of the routine work required to sustain an efficient society can be carried out by such communication-based automation, freeing humans to do more critical tasks.

Indeed, many of the expected advances of the next decade require some form of automatic operation. Fourth generation mobile devices, wearables, pervasive computing etc. have a requirement that many of the interactions between devices take place without undue demand on the user. Put another way, they need a form of self-configuring and distributed operating system that is capable of a dynamic "plug and play" functionality that takes into account social priorities. Other potential applications are only limited by our imagination, but they all have a potential social dimension such that the systems need to have a form of *social knowledge* built into them. Research workers are already giving more and more attention to encoding social intelligence, in which it is possible to express matters such as rights, obligations, permissions and authorities, leading to issues of sanctions following any form of misuse of automation.

The idea of any kind of automation is that we come to rely on it and we learn to take its functioning for granted. We cannot say how *machine-to-machine* communications will be brought about. The standards body called FIPA is creating standards that can be used for developing software agent-based services. It is uncertain, however, if we will actually see a successful launch of such services in the near future. A plausible alternative scenario is that we will continue to relegate more and more of the routine tasks to automatic transactions as we gain confidence in the way they operate, such that after a few years we will come to realise that a large amount of traffic on communications links is actually from one machine to another without any direct intervention by human users. Such a radical form of automation is bound to have pathologies, some that we can imagine in advance but many that may not be foreseeable. As long as profitable applications can be built using automated transactions between software entities, we are bound to see an increasing use of them, whether we continue to call them software agents or not.

1.4 Lessons Being Learnt

Three of the most important recent phenomena of convergence have been the World Wide Web, the increasing use of email, and the mobile phone. The growth in the first two has been incremental, with businesses as well as standard bodies working hard to keep pace with the demands of the users. Mobile phone use has required a significant amount of initial investment in the infrastructure before the services can be offered seriously. We need to pause to admire the technological innovation and drive that has gone into making mobile telephony what it is today. The heavy handsets of the first generation mobile phones of only a few years ago have been replaced by modern handsets that pack in an enormous degree of both digital and analogue circuitry, whose sophistication would have been unthinkable 20 years ago. This is in fact a techno–economic success story because it has needed business innovation as well.

Experience also shows that not every new technology product introduced has succeeded. Sometimes failures have more to teach us than successes. Some of the interesting failures in recent years are:

- the Iridium satellite-based global telephone system
- WAP protocol-based information for mobile handsets
- the rejection by the public of video on demand services
- the slow take-up of digital broadcast services

The Iridium concept is not short of its own excellent technological innovation. Clearly it is a mistake to think of these convergence-based products (successes as well as failures), as merely technology push phenomena. There is also a social aspect to these phenomena, leading to a considerable level of demand-pull. The real technical success story is not just in bringing about the original innovation (although we are conditioned to give all the credit to that original innovation), but to be responsive to the evolving demand requirements by allowing a rapid evolution of the technology after the initial introduction. Leading companies have learned that it is not enough to introduce new products; they have to be prepared at the outset to learn to respond to the market rapidly as it evolves. It is the demand-pull that has created economic successes like email, the World Wide Web and short messaging services. These services were never intended to be what they have now become.

Products succeed because they meet the needs of the users, the principle being that the users of the technology in question find themselves in an advantageous position compared with non-users of that technology. Coupled with that is the fact that in all communications-based services, network externalities add a multiplying effect. I use Microsoft Word for creating my documents, not because it is an outstanding technical product, but mostly because almost everyone I know uses Microsoft Word.

1.5 Information Ecosystems

At the height of the optimism about convergence some Internet guru remarked that the Internet is the most important event that has occurred in the history of mankind – "Think of any key event in the history of mankind and the Internet is bigger than that", he said. Can such a claim be substantiated? Another similar statement was that since the Information Ecosystem is such an important part of human life on this planet, the digitisation of it would inevitably have a significant impact. (The philosopher Sir Karl Popper has many essays on what he calls World 3. Just as a spider creates a web by its very existence, we humans create information – an objective world distinct from the physical world and the world of inner experience. The difference is that new generation of spiders do not as a rule take advantage of the webs created by their ancestors, but we humans do build upon the information that has been created by our ancestors. The Information Ecosystem is a similar notion and it is the key to the continuing progress of our civilisation.)

We need to stand back from the current euphoria to examine these claims. The actual digital age is only about 50 years old. It appeared at the tail end of the last millennium, but it will reach maturity in the new millennium. How will it shape the coming millennium? Perhaps we should ask, what were the significant things that shaped the previous two millennia? Our own present culture, with its scientific outlook, which was in the end capable of creating the digital revolution, occurred in the last 300 years or so. Can we say that the printing press had a significant effect on the promotion of scientific enquiry? Note that the printing press itself did not help to create new information; it merely provided easier access to information and perhaps as a result had a multiplying effect on the creation of new information.

The invention of writing itself is almost 7000 years old. It may be possible to say that the invention of writing was a side effect of an agrarian society based on the invention of agriculture. The separation of the production of foodstuff and its eventual consumption required some form of accounting of stocks. The origins of writing may have been principally for account keeping and then subsequently adapted for story-telling, both factual and fictional – but mostly somewhere in-between. Can the Internet really outperform such significant events in our history? Is the discovery of penicillin (and indeed other medical advances) perhaps responsible for reducing the role of religion in our lives? Is the birth control pill responsible for promoting equality between men and women? Is air travel rather than the Internet responsible for bringing down barriers between different cultures? Note that many already claim that the Internet is creating a global society. This may be true as far as businesses are concerned, but culturally the credit has to go to easier air travel. Surely these are vital questions to ask if we are to assess the true impact of the convergence-related technologies.

Incidentally, it is obvious that the television has had a significant impact on family life. What the digitisation of it will add to life is not that clear yet. At present we merely have a digital version of the familiar analogue television without any additional truly interactive possibilities that can be exploited within a digital infrastructure. One way to evaluate the impact of new media technologies is to examine the various elements in the value chain between the creation of content and its ultimate consumption. The main interaction with digital technology occurs at the two extremes. Authoring tools are expensive and can only be used by expert intermediaries. All the viewing and rendering is on a glass screen which has severe limitations – principally the number of pixels on a screen (i.e. the resolution) and the dynamic range of contrast on the screen compared to paper; and this is not to mention the weight of it and the need to provide it with power.

1.6 Conclusions

In conclusion, we should take note of the fact that computer networks in partic-ular have exhibited a clear trend that intelligence is located (or decision-making takes place) near the edge of the network and as close to the human as is possible. This is simply because Moore's Law has allowed us to locate a consid-erable amount of computational power in digital devices. The other networks – telephone networks and television networks – have tended to keep decision-making as centralised as possible. That is a mind-set problem inherited from the analogue legacy of those two areas. Even there, intelligence has often managed to leak out towards the edges of the network – consider the telephone that plays different dial tones based on the person calling a handset and based on information contained in the local directory. We may therefore be able to assert that in all areas of convergence we will tend to see this feature of decision-making tending to take place near to the human user – at first as a feature belonging to the device that users can configure for themselves, and later on as a feature that is invoked automatically.

Overall, digital technology appears to be less friendly than what has gone before. We are analogue beings who prefer to act intuitively with the world around us, whereas the digital technology demands a large amount of cognitive action on our part. But then that itself may be a false analysis. The real users of this technology are only just being born. The devices we use have an enormous scope of further improvements and research into new methods of interacting with machines is still in its infancy. Furthermore, we humans have an infinite capacity to encompass a skill that demanded cognitive action at first and turned it into an everyday intuitive skill within a very few generations.

There is a real possibility that we will see an enormous amount of machine-to-machine automation using the communications infrastructure to help in sustaining the normal day-to-day life of us humans. This automation will need the devices to possess a considerable amount of social intelligence. In the end the carbon life form (i.e. us), and silicon life form (i.e. intelligent machines) may have a profitable symbiotic relationship.

2

Interface Issues for Wearable and Mobile Computer Users

R. S. Kalawsky

Abstract

This chapter describes an important sketch-based user interface for in-field wearable/mobile computing platforms. In this context, the term "mobile computing" is taken to mean in-field mobility, where users are not seated at a desk or resting comfortably with the computer on their lap. The in-field computer user is someone who is required to enter data whilst on the move (such as walking around) or in a difficult situation (working outdoors) where a conventional laptop-type computer would be impractical. Conventional software applications, such as Microsoft Word do not readily map onto these types of user. The required degree of precise control input would be too demanding whilst on the move. This chapter deals with the requirement to enter complex information that has a spatial context and will present a sketch input system developed at Loughborough for spatial information entry by a highly mobile worker. The sketch interface is capable of supporting military operations, police, accident investigation teams, scene of crime data gathering etc.

2.1 The Wearable and Mobile Computer

The development of very small portable computers has enabled users to operate computers away from their desks and use them whilst on the move: on the train or in locations of their choice. Such users tend to operate their portable devices whilst seated at desk or on their laps, and in this sense their workplace remains static. However, there are other users, such as scientific field workers, emergency services, surveillance teams and military personnel, who require computing on the move. This category of portable computing is known as mobile or wearable computing. The term "mobile" refers to in-field mobility rather than reference to a laptop computer being used remotely from the office. Consequently, a wearable computer refers to a device that is worn in some way. As the concept "wearable" suggests, the aim is to bring computational power closer to the user so that, ultimately, they form one and the same system: physically and functionally. However, these are much more than just portable

8

computing systems: they offer the potential for fully integrated interfaces which aim to synthesise the user and task, through seamless interface technology, to greater levels of operational capability. As such, wearable computers may offer dynamic and flexible alternatives to conventional systems design and the potential for more sophisticated command, control and communications systems. In order for these devices to be effective in the field they must be small, as well as easy to operate in quite difficult or demanding working environments. It is conceivable that some wearable computer users will need to crawl, walk or stand whilst using their systems. The use of a conventional keyboard and mouse would under most circumstances be totally impractical.

2.2 Distinguishing Features of Mobile and Wearable Computers

There are distinct feature differences between mobile and wearable computers, so that the associated user interface is very different. Mobile computers tend to be small devices that enable the user to stow them in a pocket whilst not in use. When required they can be pulled from the pocket and switched on in order to enter data or simply view previously stored information. Rhodes (1997) offers a comprehensive definition for wearable computers, stating five main characteristics:

- Portable while operational – the most distinguishing feature of a wearable is that it can be used whilst moving around. This distinguishes wearables from both desktop and laptop computers.
- Hands-free use – military and industrial applications for wearable computers emphasise this point and concentrate on head-up displays (HUDs), direct voice input (DVI) or direct voice output (DVO) devices. Other input devices may include: gesture recognition, chordic-keyboards, dials and joysticks to minimise hand movements.
- Sensors – in addition to user inputs, a wearable should have sensors for the physical environment, such as wireless communications, Global Positioning Systems (GPS), cameras and microphones.
- "Proactive" – a wearable should be able to convey information to its user even when it is not actively being used. For example, if a message comes through a communications channel, the interface should be able to communicate this information to the user in some manner.
- Always on and running – by default a wearable is always on: working, sensing and acting. This is opposed to the normal use of computers and pen-based handheld systems, which normally only run when a task needs to be carried out.

2.2.1 The Wearable and Mobile Computer User's Operational Environment

Wearable/mobile computers have the potential to change the physical nature of the interface, and this challenges current user interaction styles. With a conventional PC and standard office application, the user's primary task is to concentrate upon the computing task in hand, for example word processing or data manipulation. Wearable/mobile computers differ because they take on more of a supporting role while the user undertakes another, possibly uncon-nected, task. Therefore the computer must support the user by providing infor-mation based on the current context of the working environment and task in hand to allow users to act upon this information as and when they feel it neces-sary. Wearable/mobile computers also differ from more conventional systems because their users operate in a 3D spatial environment rather than being confined to a static situation.

2.3 Interfaces for Wearable/Mobile Computer Users

2.3.1 Input Interfaces

Many user interface technologies are appropriate for mobile users; some are very sophisticated and others are little more than miniature versions of desktop systems. When used in a traditional desktop mode they usually work well, but as soon as the desktop disappears they become difficult or impossible to use. Rather than attempt to describe every available user interface for a wear-able/mobile computer it is best to identify the main systems, as shown in Tables 2.1 and 2.2.

The input devices associated with wearable computing are wide-ranging in nature. In many applications, direct voice input (DVI) is an appropriate means

Table 2.1 Input interfaces.

Technology	Wearable computer	Mobile computer (e.g. PDA)
Keyboard	OK – worn on arm, may be impossible to use operationally. Real or fabric-based keyboard	Ok – on screen keyboard. Separate keyboard
Chordic keyboard	OK – hand-operated/carried	Not very useful
Dedicated function keyboard	Excellent – small number of functions provided. Can be attached to other personal carried devices	OK to a point, but inflexible
Speech recognition	OK for hands-free interaction – note social implications when used in pres-ence of other users. Reliability outdoors is very problematical	Possible, but limited application until technology improves
Mouse/trackball	OK – a visual display must be used. Requires fine motor control which may not be achievable whilst walking or taking part in other activities	Not really suitable. Larger format systems OK – visual display required
Pen input	Non-hands-free, but fast selection. OK for drawing, Poor for significant text input (Pascoe et al., 1998)	Excellent for speed selection
Eye tracking	Expensive – immature technology	Too expensive and impractical

Table 2.2 Output Interfaces.

Technology	Wearable computer	Mobile computer (e.g. PDA)
Head-mounted display	OK – needs very careful integration with user and external environment	OK – not clear where the real benefits are
Visually coupled system	Excellent for spatial awareness. Image registration with real world problematical	Unsuitable
Flat panel display	OK – operational impact	Excellent
Touch screen	OK for menu selection or pen input. Alternative to a mouse	Ideal when used with pen input devices
Auditory display	Excellent hands-free potential	OK – limited potential

of control. Indeed, a completely hands-free approach is particularly suited to the desired needs. However, the requirement for diagrammatic expression supersedes this factor. The effectiveness of some wearable computer control technologies is assessed in McMillan *et al.* (1999).

Input methods can also include fingertip cursor and pushbutton controls. A popular system for early generation wearable computer applications is the Handykey "Twiddler" device (Figure 2.1), which is classed as a "key-sequence" input device. The Twiddler is a one-handed device, which combines keyboard and mouse functions in one device to provide textual input and browsing via a cursor. The manufacturers claim that a trained user can type up to 60 words per minute using the Twiddler's chordic keyboard system, which makes the wearable computer suitable for extended data entry. However, not all users agree. The index finger-driven mouse enables the user to navigate a GUI environment just like any conventional windowing environment, but is very challenging for novice users.

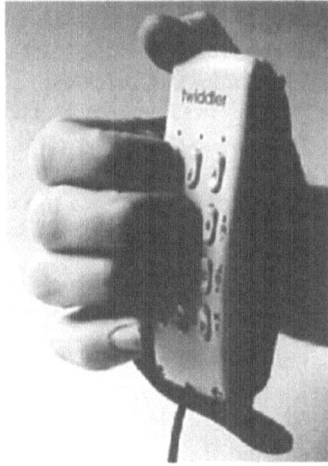

Figure 2.1 Handykey Corporation's "Twiddler".

Pen input systems are becoming very popular with the advent of Palm Pilots and Windows CE-enabled handheld computers. These rely on exerting location-sensitive pressure on a flat, physical touch screen. Conventional touch screen technology often relies on "stylus" utilisation – basically a pen-size pointer. However, this introduces other context-sensitive issues: where is the stylus to be stored when not performing information entry tasks?; how long does it take to obtain it from this storage point?; is the system still operable if the stylus is broken/mislaid?; etc.

2.3.2 Hands-free Technologies

A good wearable computer should facilitate hands-free interaction so that information is presented to the user only when required. Speech recognition would at first appear to be a good method of controlling a wearable computer. This may be true for environments where the ambient noise level is fairly low. However, if the user is working in an area where sound levels are unpredictable or very loud, speech input may be ineffective. Additionally, from a social perspective the use of speech input can be problematical in certain situations. Consider the ethics of speaking out loud the medical symptoms in front of a semi-conscious patient.

In order to provide automatic effective hands-free input the wearable/mobile computer must ideally track what the user is doing whilst being aware of the context of the surroundings. These requirements immediately place a big demand on the sensor technologies and the processing used by the wearable/mobile computer.

2.3.3 Output Interfaces

A mobile computer is by definition a small device, and this implies display technology of a reduced size. Small screen sizes and low display resolutions make the design of an associated user interface difficult – display "real estate" is a critical factor, along with the legibility of symbology. For these reasons, the larger the screen dimensions and resolution the better. The display area within which to input and display symbology information is to be maximised, whilst still providing intuitive control over system behaviour and functionality, aimed to lower cognitive workload in the field (a critical factor in this context). In the event of mistakes being made during information input, attributed to this cognitive strain, provision must be made for the ability to "backtrack" the input procedure and effectively "undo" mistakes made.

2.3.4 Visually Coupled Displays

A visually coupled display provides visual display information in the line of sight of the user. This display mode requires a head-mounted display to superimpose imagery onto the real world. By providing important information in the user's line of sight they do not need to take their focus of attention away from the task in hand. The use of virtual display overlays has been extremely successful in military cockpit systems, where the head-up display provides this information overlay. See-through head-mounted displays are also appearing in certain military vehicle applications. A notable characteristic of these head-mounted displays is that the displayed information is displayed at infinity, thus eliminating the parallax error when viewing distant objects. This also removes the need for pilots to re-accommodate their eyes between viewing the objects in the outside world and when they look at the display. However, it is impossible to fuse the head-mounted display with information that is very close – the disparity is simply too great. Unfortunately, many non-military head-mounted display applications apparently ignore this and require the user to accommodate on both distant and near-field information whilst viewing display symbology projected by the head-mounted display. To alternate between outside world information located in the near and far fields would require dynamic optical correction – a very expensive option.

A visually coupled display requires head orientation information in order to relate the displayed information to real-world coordinates. Conventional virtual reality head-tracking systems are not practical propositions for mobile applications because they require a fixed reference transmitter unit. In most cases there are practical difficulties in setting up electromagnetic or optical position transducers. The approach taken at Loughborough is to employ lightweight fluxgate magnetometers and fluid-based sensors to detect head orientation. Even these do not yield the required resolution for accurate registration of computer-generated symbology onto the real-world imagery. The registration of computer-generated imagery onto the outside world is arguably one of the most difficult challenges for wearable computer applications. The errors are particularly noticeable whenever a computer image is overlaid onto an object that is positioned close to the user. Design solutions round this problem are not discussed in this chapter.

2.3.5 Display Formats for Wearable/Mobile Computer Users

There seems to be an unjustified adoption of Microsoft Windows-style metaphors for many wearable/mobile computer applications. Whilst the commonality with the desktop systems seems attractive, it is possible that this approach may not be the best for the user. For head-mounted display applications, the user is presented with a serious problem when it comes to making selections from the display. Control of a windows, icons, menus and pointers (WIMP) interface requires a particular skill that entails fine motor control to adjust a cursor in a continuous manner across the screen, often to pixel accuracy. Users must keep their eyes fixated on the cursor at all times during the selection process. Movement of a cursor or other marker to indicate a point of interest is feasible, but the need to perform word processing or spreadsheet type text input tasks via an HMD is problematical. The whole purpose of HMDs is to augment the user's view of the real world with additional computer-generated imagery. This suggests that the user is concentrating on a primary task and the computer is required to provide supplemental information that might aid that task. If the display is moving as well as the user then pointing with pixel accuracy is extremely difficult, to say the least. This method of interaction is totally inappropriate for HMD display formats. In see-through HMD applications it is possible for the display symbology to completely obscure the point of interest. This is a serious dilemma for the designers of display formats used for wearable/mobile computing systems.

The display/control paradigm of a wearable computer does not lend itself to a direct manipulation interface, unlike desktop applications where the process of inputting data becomes the primary task. It is argued that this aspect of user attention is one that differentiates a wearable/mobile computer from a laptop or desktop situation. Unless HMD-based wearable computer displays are carefully designed the user can be put under an unacceptably high cognitive load that interferes with the task in hand (Stedmon *et al.*, 1999b; Kalawsky *et al.* 2000a,b). The wearable computer, by virtue of being a secondary system, should present a low cognitive load to its user. It would seem to make sense to incorporate some form of intelligent agent in order to process information and only make this available when the user requires it or in order to ensure that the user maintains a state of awareness of what is going on. With the right level of design it is possible to present the user with a very simple interface – accept or reject.

2.3.6 Auditory Output

Auditory output is perfectly adequate for some mobile computing applications. As Brewster *et al.* (1998) suggest, sound can improve interaction and may prove useful in limited display devices. Brewster *et al.* (1998) also concluded that non-speech sounds (in combination with speech) can help users navigate through menus of options more easily. "Earcons" or auditory icons have been used in military applications for many years to great effect.

2.3.7 Which Interaction Styles/Metaphors Apply Best to the Wearable?

Traditional desktop computer metaphors are not appropriate when describing a system that mingles so closely with the human body. Users are not used to the concept of having continuous system access as intimate as that offered by, say, a wristwatch. Therefore a different set of metaphors will be required for a wearable system. The context of use will influence the design of the wearable computer's input/output interface. For example, systems that rely entirely on head-coupled information will employ graphical symbology that can be read against a background scene that is transmitted through a head-mounted display. The need for detailed requirements capture and task analysis cannot be emphasised enough, because unlike a desktop application the wearable computer user is often multi-processing between different tasks.

2.4 A Novel User Interface for a Wearable Computer

This section describes the sketch input user interface that has been incorporated into a wearable computer system at Loughborough University for military applications. The emergent system has proven to be a powerful method for spatial expression input, which is considered to be difficult to achieve with other user interface whilst on the move. A user task analysis had to be performed to identify the critical design parameters of the interface in order to design an appropriate solution. The context of use of the iFC needs to be understood before the sketch input mode is described.

2.4.1 The Loughborough iFC and Its Operational Environment

The operational environment in which the iFC is operated needs to be reviewed to understand the context-specific considerations. The context in which the iFC is to be deployed is the military domain, particularly with a view to equipping infantrymen in a battlefield scenario. The study by Stedmon *et al.* (1999a) identified the key issues. The degree of hostility present in this type of operational environment has implications for the detailed design of the system and its operational characteristics. For example, electronic-based hardware worn on the body could be susceptible to damage in the field (whether as a result of combat or of rough treatment in the physical vicinity). Inherently, system inoperability could expose the user to a greater threat in the field, due to downgraded fundamental operational capability. Operations undertaken in the battlefield may have unpredictable durations – a sortie could last from minutes to days, and an adequate power supply must be provided throughout the duration. Therefore battery life and recharge time (or system "turnaround time") must be considered. Another critical factor is that the system should be operable at short notice (i.e. come out of its "idle" state, as quickly as possible). The system design should account for exposure to "the elements" – this includes extreme temperature levels, moisture and precipitation, and low/high ambient light levels. These factors affect technical aspects such as operating temperature ranges, water resistance and display/control luminescence, respectively. The hostile environment also affects the user interface design, so it is important the system does not compromise user effectiveness or personal safety in the field. The latest iFC is shown in Figure 2.2.

Figure 2.2 Side view of Loughborough iFC.

2.4.2 The iFC and Its Operational Context

The iFC has been designed for military in-field applications where the user requires limited or continuous support from the computer. User tasks include accessing secure databases, navigational information, tactical data, control of system, communications support (audible and multimedia), input of command and control data and intelligence data input/output. The diversity of information requires a multi-modal control/display interface to suit the needs of the user at different times during a mission. Different operational environments make certain control/display modalities impractical for certain tasks. For example, the operational environment may change in such a way that a variety of user interface modalities must be provided. Consequently, a variety of input methods have been incorporated and range from conventional keyboard input, chordic keyboard input, and pen input through to speech recognition. The iFC must be sufficiently adaptable to accommodate multi-modal input across changing task demands. A detailed task analysis was undertaken and important requirements emerged. One very important requirement for information input emerged which was concerned with rapid tactical data input that could subsequently be transmitted to other users in the area. After receipt of this tactical data the users would be required to amend or update the tactical representation and send this back to the originator.

2.4.3 IFC Rationale for an Alternative Non-keyboard Input Method

The mobility of the in-field military user precludes the use of a keyboard for many operations. The conventional qwerty keyboard occupies a relatively large surface area – too large for it to be operated in our mobile situation. Even though miniature keyboards have been developed which can be worn on the arm, these tend to get in the way (see Figure 2.3). The issue is not so much one of robustness and reliability, instead the key question was the efficiency and time of text input to describe complex situations in an unambiguous manner. Whilst text input using a keyboard is trivial, expressing command and control data for a complex battlefield scenario in a very clear manner is problematical.

Figure 2.3 Wrist PC keyboard.

2.4.4 Expression of Spatially Based Information

Certain in-field applications make use of spatial information that relates to the position of objects in the environment. This can include the specification of an object's position via latitude and longitude or its relationship to other objects in the environment (e.g. overlays on map displays). The obvious way of detailing this spatially based information is through the use of pictorial displays. This information is best presented to the user as a tactical overlay on the real world or perhaps as a symbolic representation on a tactical map display – the old adage "A picture paints a thousand words" is entirely appropriate. A further requirement is the need to avoid display clutter or obscuration.

The information input format needs to be addressed as well as the method of presentation. Because the battlefield is not simply a flat surface on which notable entities lie, one could ask the question: should expressed information be in a format other than 2D top-down? There have been numerous studies to analyse the basic qualities and capabilities of two- and three-dimensional views. One such study, "Using 2-D vs. 3-D Displays: Gestalt and Precision Tasks" (St John and Cowen, 1999), concludes that whilst spatial 3D views are better for 3D-based shape understanding, 2D representations are far better for precisely judging relative positions of objects. Therefore the adoption of one or the other should very much depend on the information to be expressed. The nature of standardised symbology in this context is of a top-down nature, and hence this imposes a 2D constraint on the display type. For instance, if users want to express their line-of-sight view, the symbology convention is likely to prevent this possibility. Although a top-down approach has to be adopted, these studies could prove beneficial if a consequent 3D view were to be developed at a later time.

Exchange of audio-based information in a mobile team situation (e.g. remote communication between infantry) may not be sufficient to effectively translate this information to the rest of the group – information may be misheard, forgotten (temporally based communication), overly verbose etc. Whilst auditory expression is fast it can be regarded as an unreliable or inefficient communication method. An alternative to auditory information exchange is the use of a text-based system – through a keyboard and mouse. The problems associated with this input modality for highly mobile users have already been discussed. However, consider the extract from a tactical message given in Figure 2.4.

Passing this message over an audio communications link could lead to errors, especially if the message had to be passed onto others in the group. Even though a typewritten message such as this would only take a short time to type on a

Fire Team A will attack from the west side of the woodland whilst Fire Team B will circle round to the north of the hill. Intelligence suggests that the enemy AA position is defended by approximately 20 troops and that a machine gun emplacement is situated at the south west corner.

Figure 2.4 Textual expression of orders.

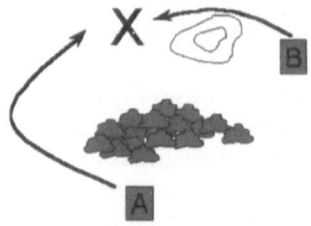

Figure 2.5 Diagrammatic expression of orders.

conventional keyboard, the use of a keyboard may be impractical. However, textual expression can lead to ambiguous information and errors if a lengthy message is composed very quickly.

A more effective way to communicate this information to others would be to express the situation in the form of a diagram (Figure 2.5), incorporating a standard symbology and notation standards. For instance, the translation of the tactical message extract could be achieved by "sketching" the situation, utilising known symbol sets and established notation standards, into a touch-screen-based PDA. This "sketch" input could then be transmitted and output to similar screens, "worn or carried" by other group members.
The transmitted "sketch" could be passed to the group leader for further augmentation, and retransmitted to express a subsequent battle plan for the group, with respect to this new shared information. This form of communication allows for the expression of spatially based information and scale, and when combined with the application of an established battlefield symbol set (US DoD, 1996) can comprehensively present a large array of information with ease.

2.4.5 Integrated Perspectives

In certain situations a group of people can be working on the same spatial problem and see the same information from different angles or perspectives. If spatial information between these groups were to be passed between each other as text it is possible that the correct perspective would be lost. The problem can arise from a lack of a baseline or reference. A diagram does not suffer from the same problem if a clear anchor is used, such as a landmark or geographic feature. Consider, for example, a command and control installation that could be viewing the digital battlefield status, thanks to transmitted data from a variety of friendly units and intelligence sources. However, details of enemy unit types and positions may not be known in a certain area. Infantry groups located within and around this vicinity could input what they perceive to be their perspective on the area in question, in a diagrammatical fashion. If a number of these "perspectives" are received from a number of distributed infantrymen, then the multi-sourced information can be integrated to give a more definitive view for use in command and control. Figure 2.6 illustrates this concept.

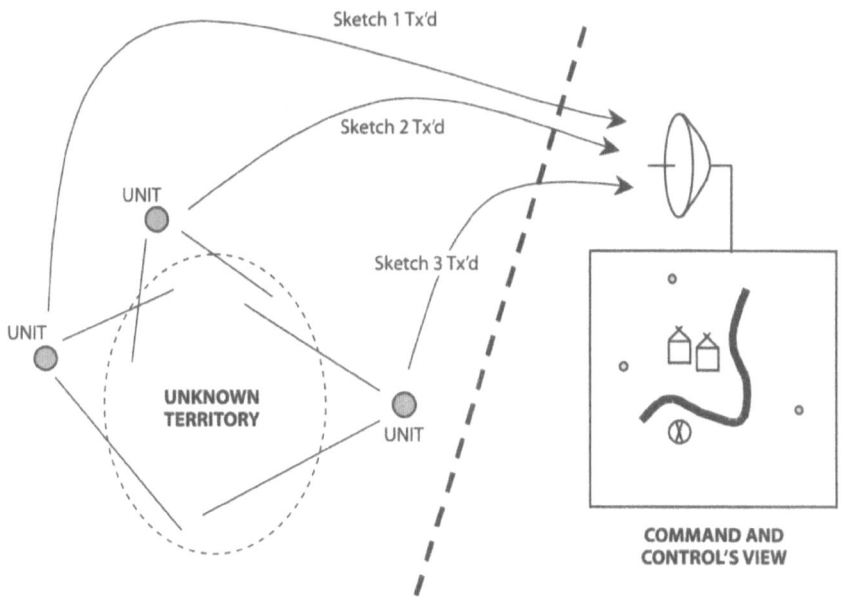

Figure 2.6 Integrated "perspectives".

An array of equipped infantrymen may each collectively receive a request, via the concept system, to send back geographical or enemy information about a given area. The different units proceed with inputting their subjective "knowledge" diagrammatically, and then selecting a control to subsequently transmit the graphics back to the requester.

A variation of this scenario was also envisaged. For example, a battle group might be planning a move into a given vicinity. The intelligence information associated with this area might be somewhat sparse. For instance, it may be known that enemy forces are present, but the magnitude, type and location of those forces are not known. The battle group could "send out" groups of system-equipped infantrymen to penetrate the area more subtly, and report back remotely to the group on their gathered knowledge. This concept is illustrated in Figure 2.7.

This collective information is received by the main group, and a clearer picture of the area in question is developed. This can be analysed so the battle group's task of moving into the area is likely to be significantly more successful.

Another example of the utilisation of the concept system differs in nature somewhat. A system-equipped soldier may be prompted by a command and control role to report on the current status of friendly/enemy units in his or her vicinity. This could be done by manually annotating a representative map of the area in question, which is transmitted to the system with appropriate notations/symbols. This would be followed by the return transmission of the annotated map graphic to the command and control role.

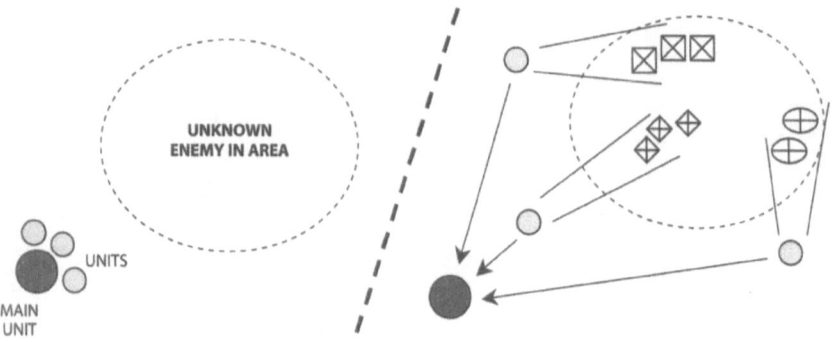

Figure 2.7 Example of reconnaissance-type utilisation.

The concept system could also act simply to dynamically enhance the infantry-man's current situation awareness. The infantryman could utilise the system to, on request, "download" up-to-the-minute graphic information, expressing current unit positions, status and other associative information, from a command and control source. Control over the *type* of information displayed could also be provided by the system. The infantryman could switch between displaying purely mobile units, fixed installations, geographical detail etc. – all information sourced from the command and control role. An active communications "channel" could even be set up to suitably alter (animate) the view to reflect changes in the source digital representation of the battlefield, as and when it occurs. For example, a group of enemy tank units may begin to move off in a given direction. If they are being tracked, then this positional data will, first of all, be fed back to the command and control role, whereupon the digital representation of the battlefield will be being updated. This will then be reflected on the infantryman's system as new positional data is actively "streamed" in.

2.5 iFC Sketch Input System

A decision was taken to explore the use of a sketch mode system for the iFC as an option for rapid tactical data entry. The motivation was a desire for a system that would exploit the expressive capability of sketches and the powerful transfer of information that a picture can convey. In the military domain it is commonplace to employ symbolic representation on maps and other display output to express a tactical situation. Therefore it was considered to be a rela-tively small step to implement a sketch input system.

2.5.1 Symbology Study

Standardised symbol sets currently exist for battlefield representation, and are applied in military force coalitions around the world. It has been reported (US DoD, 1996) that "Standard symbology synthesized from land-based, nautical, and aeronautical warfighting domains is an increasingly essential ingredient in the successful implementation of the Command, Control, Communications, Computers, and Intelligence for the Warrior (C4IFTW) concept". One such symbol set is the Common Warfighting Symbology (US DoD, 1996).

The joint operations undertaken by these forces bring about an increased requirement for a rapid exchange of information between groups in the cross-community. These information transactions invariably require some element of standardisation, so a joint symbol set, eliminating conflicts between the various groups' scenario representation practices, was developed. The standard builds on human factors and cognitive ergonomics research, and has been "moulded" through users' input, ensuring it meets the warfighters' require-ments. It includes the symbol hierarchy, information taxonomy and symbol identifiers that should be adopted in the transaction (electronically or other-wise) of information in an operation's "infosphere", and is stated as being applicable to both automated and hand-drawn representation.

The extensive documentation for the Common Warfighting Symbology was analysed, and provided an indication as to the typical characteristics of a meaningful inputted "sketch" and the symbolic components that it may encom-pass. The iFC sketch input system was required to support such expression, albeit in a scaled-down form.

2.5.2 Warfighting Symbology Composition

There are two definable discrete types of symbology: "icon-based" and "tactical graphics". In conjunction they constitute the expressive requirements of the "warfighter" active in a number of wide-ranging scenarios.

Icon-based Symbology

The icon-based symbology is designed for effective representation of objects, e.g. tank units. It indicates an object's affiliation (unknown, friend, neutral and hostile), through variation of its shape and colour. It also portrays the object's "battle dimension" – its primary mission area. This can be above, on, or below the Earth's surface, and is distinguished by manipulation of the icon's generic shape.

Tactical Graphics Symbology

This symbology provides operational information that cannot be presented via icon-based methods alone. Typical portrayals are boundaries, area designations,

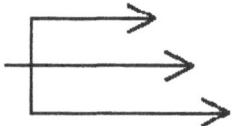

Figure 2.8 "Disrupt" symbol.

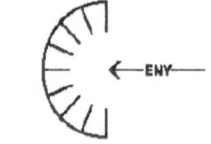

Figure 2.9 "Contain" symbol.

and other unique markings corresponding to battlefield dynamics, planning and management. More specifically, the tactical graphics allow expression of tasks, manoeuvres, reference points, areas, deception techniques, fire support strategy and other command and control information. It has been designed to be applied alongside icon-based expression. Tactical graphics tend not to rely on use of colour – the default colour is either black or white, depending on the background colour.

This symbology accounts for a vast array of situation and scenario representations. Its diversity makes it applicable for virtually every conceivable scenario that a warfighter may wish to express, or indeed, interpret. The complete detail and definition of this symbology is included in Appendix E of US DoD (1996). Some examples from this are included below.

Figure 2.8 depicts the symbol for a disruptive tactic – "A tactical task or obstacle effect... that breaks apart an enemy's formation and tempo, interrupts the enemy's time table, causes premature commitment of forces, and/or splinters their attack". The symbol would be applied (spatially) to a given displayed situation, and would be in the context of accompanying icon-based symbols. Expressions of this nature may need to be complemented with textual information. Subsequent figures demonstrate this.

Figure 2.9 depicts the symbol for a containing tactic – "To stop, hold, or surround the forces of the enemy or to cause the enemy to centre activity on a given front and to prevent his withdrawing any part of his forces for use elsewhere...".

These examples demonstrate the expression of a procedure or tactic to be applied. Symbols can also be applied to express areas or zones spatially in a given context. Figure 2.10 demonstrates this. This symbol indicates "An area the commander designates as restricted from the emplacement of man-made obstacles, normally to facilitate future operations".

Figure 2.10 Zoned area.

272100Z SEP - 300400Z SEP

Figure 2.11 Obstacle-free area.

There are a multitude of area-type expressions, from "decoy mined" areas to allowable zones of engagement. The symbol set also varies widely in its levels of complexity and intricacy; refer to Figure 2.11.

Although the symbol set is standardised for multi-force operations, and includes representations for space/air-based, land/sea surface-based, and submerged entities, it seemed that a good starting point, given that the system had been geared toward the infantryman, would be to provide exhibitory ground-based representations. Examples of such are illustrated in Figure 2.12.

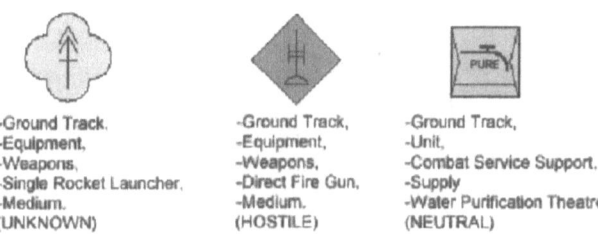

-Ground Track.
-Equipment,
-Weapons,
-Single Rocket Launcher,
-Medium.
(UNKNOWN)

-Ground Track,
-Equipment,
-Weapons,
-Direct Fire Gun,
-Medium.
(HOSTILE)

-Ground Track,
-Unit,
-Combat Service Support.
-Supply
-Water Purification Theatre.
(NEUTRAL)

Figure 2.12 Ground-based icon examples.

Some symbols can be thought of as consisting of a subset of more basic shapes and notations, brought together to express "compound" information. The issue of notation intricacy does, again, have significance for system design.

2.6 BatSAT (Battlefield Situation Awareness Tool)

The iFC application that was designed to support the expression of MIL-STD-2525 tactical graphics and icon-based symbology (US DoD, 1996) electronically via pen-based input methods is known as BatSAT. The system demonstrates the ability to share electronic information across platforms, and interact with it fully. BatSAT's operation can be broken down into three distinct components:

- The control user interface
- Information entry
- Information exchange

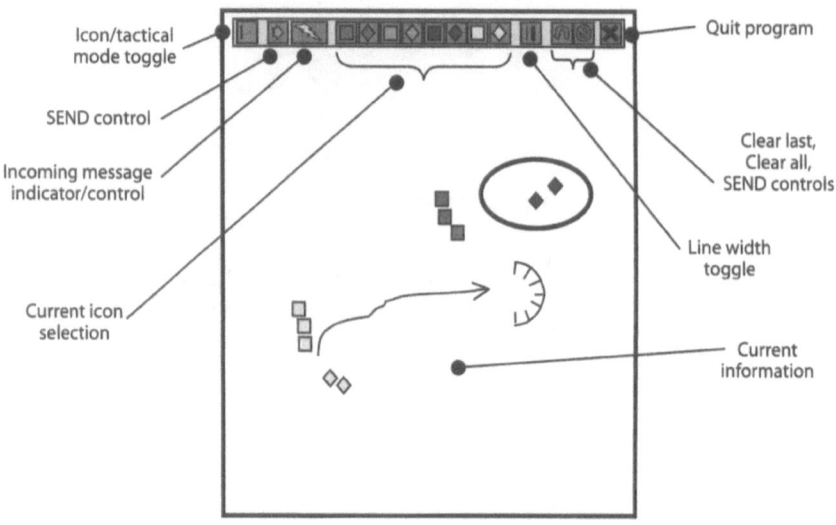

Figure 2.13 The BatSAT.

A small handheld computer was selected that handles reliable pen input and could be expanded through the use of dedicated communication links (serial, infrared and wireless). A further requirement was the need for night-time and sunlight readability.

The developed symbology consisted of a combination of lines, colour fills and textual information. A "sketched" input can easily cater for the first of these, but provision for the latter two expressions is somewhat more taxing. The ability to "fill" a manually created polygon with colour could imply complex algorithmic procedures in code, and would effectively allow the user to be creating arbitrary non-defined icons. An alternative to this would be to provide the ability to "place" pre-defined icons in the sketch. A diagrammatic view of the completed BatSAT system is shown in Figure 2.13.

The main component of the user interface (Figure 2.14) takes the form of nine control buttons, located at the periphery of the main information entry and display window. The main control buttons are described below.

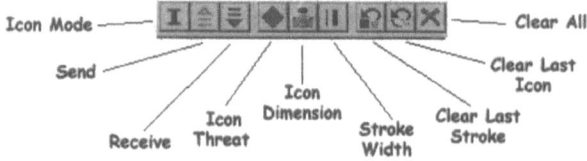

Figure 2.14 The control user interface.

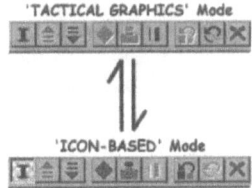

Figure 2.15 "Icon Mode" toggle button.

"Icon Mode" Toggle Button

This switches the application between its two main operable modes (Figure 2.15). When deselected, the application supports input of tactical graphics information (i.e. stroke-based). When selected, the application supports input of icon-based information.

"Send" Button

This button sends the inputted/displayed information to a currently "receiving" BatSAT or BatSAT DT node. The button raises, and the application returns to normal status, once the information has been successfully sent or the operation has "timed out" (after five seconds).

"Receive" Button

This button puts the BatSAT application into a "listening" mode, at which point any information sent from another BatSAT or BatSAT DT node is received and then displayed on screen. The button raises, and the application returns to normal status, once the information has been received or the operation has "timed out" (after five seconds).

"Icon Threat" Cycle Button

This is only available when in "Icon-Based" mode. This button cycles through the four possible threat designations that an icon to be placed can have (Figure 2.16). Its appearance indicates the current designation selection.

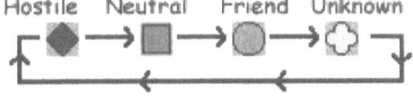

Figure 2.16 "Icon Threat" cycle button.

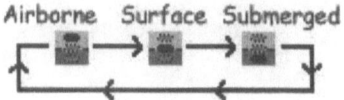

Figure 2.17 "Icon Dimension" cycle button.

"Icon Dimension" Cycle Button

This function is only available when in "Icon-Based" mode. It cycles through the three possible dimension designations that an icon to be placed can have (Figure 2.17). Its appearance indicates the current designation selection.

The full set of threat icons that can be selected by the "Icon Threat" and "Icon Dimension" cycle buttons is shown in Figure 2.18.

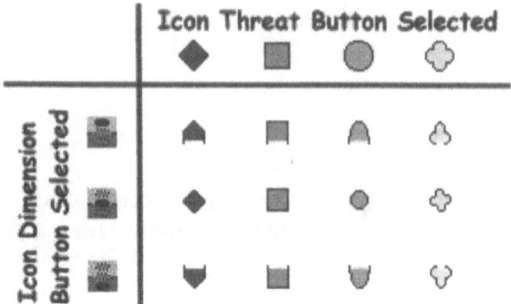

Figure 2.18 Example threat icons.

2.7 Review of System

The BatSAT system is a very efficient method for the expression of complicated spatial situations that can be entered quickly. The ability to capture information in a sketch mode is a very powerful tool for wearable computer users who are unable to operate keyboards due to their working environment. The current generation of iFC (at the time of press) supports completely wireless interaction between the handheld device and the wearable computer host. These two systems can be operated completely independently if required. Studies are in hand to evaluate the actual performance of sketch input mode compared with alternative systems for tactical data entry situations.

2.8 Concluding Remarks

The expression of complex information by sketches or diagrams is not a new concept. However, the development of such a system as an alternative for tactical handheld or wearable computers seems to have largely been neglected.

A number of application packages exist that allow handwriting or notes/ diagrams to be stored on a handheld computer, but these do not then allow subsequent editing of the information. A modified version of the BatSAT system is currently being developed for accident investigation/scene of crime operations, where this symbolic data can be linked to a remote database.

Acknowledgements

The author wishes to acknowledge Matthew Howard who coded the software environment for the BatSAT handheld computer application.

References

Brewster, S., Leplatre, G. and Crease, M. (1998) Using non-speech sounds in mobile computing devices. In *Proceedings of the First Workshop on Human Computer Interaction with Mobile Devices*, Glasgow, pp. 29–33.

Kalawsky, R. S., Hill, K., Stedmon, A. W., Cook, C. A. and Young, A. (2000a) Experimental research into human cognitive processing in an augmented reality environment for embedded training systems. *Virtual Reality*, 5 (1), 39–46.

Kalawsky, R. S., Stedmon, A. W., Hill, K. and Cook, C.A. (2000b) Old Theories, new technologies: developing guidelines for the cognitive ergonomics of augmented reality. In *Human Factors and Ergonomics Society, 44rd Annual Meeting*, San Diego, CA, pp. 3-398-393-401.

McMillan, G., Calhourn, G., Masquelier, B. L., Grigsby, S. S., Quill, L. L., Kancler, D. E. and Revels, A. R. (1999) Comparison of hands-free versus conventional wearable computer control for maintenance applications. In *Human Factors and Ergonomics Society 43rd Annual Meeting*, Houston, TX.

Pascoe, J., Ryan, N. and Morse, D. (1998) Human–computer–giraffe interaction: HCI in the field. In *Proceedings of the First Workshop on Human Computer Interaction with Mobile Devices*, Glasgow.

Rhodes, B. J. (1997) The wearable remembrance agent: a system for augmented memory. In *International Symposium on Wearable Computing (ISWC)*, pp. 123–128.

St John, M. and Cowen, B. (1999) Using 2-D vs 3-D displays: gestalt and precision tasks. In *Human Factors and Ergonomics Society 43rd Annual Meeting*, Houston, TX, pp. 1318–1322.

Stedmon, A. W., Kalawsky R. S., Moore P. M., Aung M., Purcell J., Reeh C. and York T. (1999a) It's not what you wear, it's how you wear it: human factors of wearable computers. In *Human Factors and Ergonomics Society 43rd Annual Meeting*, Houston, TX, pp. 1050–1054.

Stedmon, A. W., Kalawsky, R. S., Hill, K. and Cook, C. A. (1999b) Old theories, new technologies: cumulative clutter effects using augmented reality. In *IEE International Conference on Information Visualisation '99: International Conference on Computer Visualisation*, London, IEE.

US DoD (1996) *US Department of Defense Interface Standard Common Warfighting Symbology MIL-STD-2525A*, US Department of Defense.

3

Visual Data Navigators ("Collaboratories"): True Interactive Visualisation for the Web

Mikael Jern

Abstract

The contents in Web documents are normally restricted to static items such as text, imagery and animations. "SmartDoc" has developed "Collaboratories" (Web application components) that incorporate not only text but also the entire interactive data visualisation and navigation process into a Web document, allowing users and project teams to collaborate and share data, visualisation parameters, information and insight while distributed over the standard or mobile Internet, using intuitive visual navigation techniques. In other words, publishing a text document on the Web is only half the story. The other half is enabling others to interact with the published result and gain insight into context that is meaningful.

"SmartDoc" is a research project jointly funded by the EC Commission and focus on embedded Collaboratories that give the reader full access to any discovery and insight, data navigation tools and underlying data. Visual data navigation is provided through interactive 2D and 3D Web-based visualisation components with a small footprint. The "discovery" is described in one or several snapshots providing the history of the visualisation process. These snapshots are a copy of the component's state at the time when the snapshot was taken and allow the user to further interact from the state when the visualisation was snapped. They can be included as an image for printing the document. The underlying data or spreadsheet is either embedded in the document or accessed through a hyperlink. "What are users looking for?" is the key question guiding a SmartDoc process.

The Collaboratories (Application Components) are based an multi-layer visualisation component architecture with a small footprint suitable for Web distribution and therefore scalable and customisable to any level of expertise. A "SmartViewer" client-side plug-in, responsible for interactivity and graphics rendering, has been developed and will be distributed as "freeware" to allow free distribution of a SmartDoc on the network. Integration and assessment of application component-sharing through Web documents and a network infrastructure based on component industry standards, providing real-time data interactivity, reducing the load on the network and with zero administration client deployment.

SmartDoc scales to accommodate massive amounts of data presented in a visual format, allows full real-time interaction with on-screen presentations, and gives users an unprecedented level of high-quality visual presentation. Our integration of visualisation and data analysis through an atomic component architecture combined with special data reduction components and fast scene tree rendering by the SmartViewer enables the visual data navigation of large data sets.

3.1 Introduction to SmartDoc

The SmartDoc visual data navigation components allow the author to embed not only the content (text), but also the entire visual data navigation scenario in any electronic document. Upon receipt of the SmartDoc, the recipients can immediately share the author's insight and zoom in on specific results that affect their activities. The person viewing the report has full access to any discovery (insight), data navigation tools and underlying data. The "discovery" is described in one or several snapshots ("bookmarks") also embedded in the report. These snapshots are a copy of the control's state at the time when the snapshot was taken and allow the user to further interact from the state when the visualisation was snapped. You can visualise multivariate data and derive a deeper understanding of compound properties through data correlation and multiple display methods of the same data source. You can filter unwanted values using range sliders to expose areas of greatest importance. Isolating data of interest clears the view even further and provides close-ups of the results.

An example might be a set of multivariate data, perhaps related to product formulations, consumer behavioural studies or a set of measurement attributes. Typically a standard report on such data might start off by discussing the data in general terms and then continue with the outputs of some statistical or modelling studies to extract information from the data and then draw conclusions, perhaps coupling this with additional related information. Conceptually the interrelations within this class of dataset and between the raw data and modelling process are hard to imagine, making it difficult to extract the "learnings" from the particular problem.

Instead, imagine a smart document that takes the reader through the problem via a set of interactive visualisations coupled to the raw data. The reader can not only read the report but, if required, also do parts of the analysis from within the familiar document interface, starting with exploratory visualisation of the raw data through driving the modelling/statistical process through assimilation of the final results.

By doing this, the reader will:

- be able to explore the raw data, hopefully leading to an understanding of why a particular analysis approach was subsequently taken
- get a much more intuitive feeling for the trade-offs in the modelling approach and therefore the validity of the results
- take in the substance of the report in a very natural way.

Upon downloading of a SmartDoc to the desktop, the recipients can view text and images, but also share the author's visual insight and analysis and zoom in

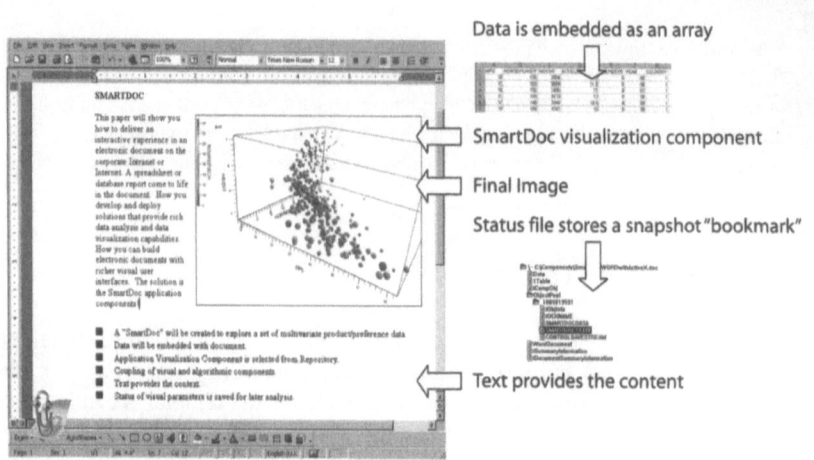

Figure 3.1 The SmartDoc represents an interactive document with embedded text (context), a visual data navigation component, data, an image for communication and the status of the exploration process.

on specific results that affect their activities. You can rearrange multivariate data to uncover patterns and trends that may go unnoticed in lengthy paper reports. A SmartDoc is based on embedded Collaboratories that give the reader full access to any discovery (insight), data navigation tools and underlying data (Figure 3.1). Visual data navigation is provided through interactive 2D and 3D Web-based visualisation components with a small footprint. The "discovery" is described in one or several snapshots providing the history of the visualisation process. These snapshots are a copy of the control's state at the time when the snapshot was taken and allow the user to further interact from the state when the visualisation was snapped. The underlying data or spreadsheet is either embedded in the document or accessed through a hyperlink.

The SmartDoc EC funded project addresses the concept of "Information Visualisation" (InfViz) through its support of:

- visual data mining techniques
- visualisation of multidimensional data
- interdisciplinary focus – visualisation, data mining, VUI, data analysis, clustering
- information visualisation techniques for large data sets
- integration of visualisation techniques in information and knowledge management systems
- combination and integration with non-visual data mining techniques
- visual data analysis and exploration for
  - knowledge management
  - consumer analysis/marketing
  - biochemical and biomedical analysis

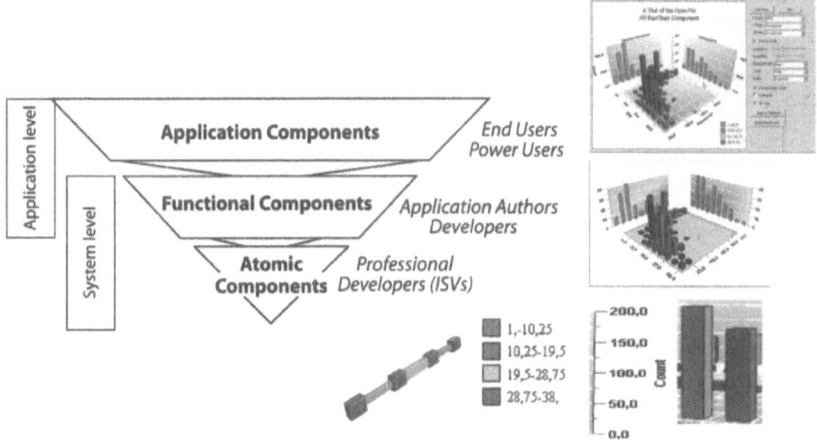

Figure 3.2 SmartDoc component mult-layer user abstraction model and system architecture.

- Internet/e-business
- emphasis on application-oriented projects

Figure 3.2 shows the SmartDoc system architecture.

3.2 Data Model

Spreadsheets have helped us organise our information into two-dimensional matrices of rows and columns. Our data model incorporates multi-dimensional data without any limitation on the number of dimensions supported. Analysis tools slice and pivot among the dimensions to display the data in any number of forms. The InfViz tools enable the user to pivot among the dimensions interactively.

There are a number of reasons for our choice of the Excel spreadsheet as the data model:

1. While we might like to think that data is managed in properly designed and maintained databases, the amount of data simply stored in Excel spreadsheets and accessible on the user's desktop is truly vast. Tools that directly address the problem of exploring this data will be of general interest.

2. While data will be drawn from a huge range of applications and problem domains, ranging from finance, IT and business management to manufacturing, research and development, there are often some common issues; for example, the data may be highly multivariate and/or time dependant. The opportunity therefore exists to develop a tool based around a particular set of visualisation techniques optimised for a particular class of data (multivariate) which nevertheless has wide applicability. Further, because

the basket of required visualisation techniques is therefore well defined, we can focus on exploring the value of the COM approach in delivering and demonstrating the performance required.

3. While the graphical techniques within Excel are improving they still remain limited to simple presentational visualisations. There are no interactive exploratory visualisation tools, so the data navigator can therefore be thought of as a tool for significantly expanding the functionality for exploring multivariate data sets available to the Excel user.

4. Excel itself is a component-based application.

3D InfViz provides several methods of viewing multiple dimensions in the same display. One way is to map the abstract data to a virtual object such as a 3D sphere graph. Five columns in an Excel spreadsheet are mapped onto X, Y, Z, Size and Colour (Figure 3.3). Effective users of this InfViz technique, however, must be analytical and multidimensional thinkers. While it seems intuitive for analytical thinkers to assess data by multiple dimensions, not everyone thinks that way. When information is organised dimensionally, as in the 3D sphere graph, the potential for understanding it grows exponentially, particularly when combined with 2D views (Figure 3.4).

The 3D sphere graph shows how various car model attributes, such as mpg, horsepower, weight, acceleration and number of cylinders, correlate with each other. The 3D spheres reveal, for example, that Japanese cars use less petrol (MPG) than American cars. Is this due to number of cylinders or weight? How is the acceleration of a car model related to its weight etc? The colleagues that you involve in this analysis will want to ask their own questions of the data. They

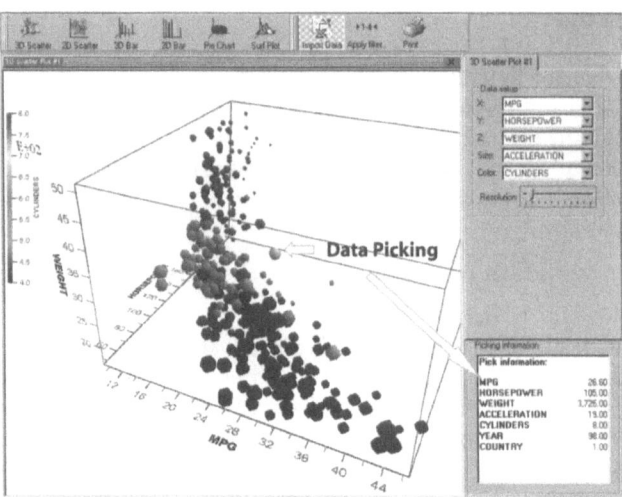

Figure 3.3 Visualisation of abstract data. Five columns (MPG, HORSEPOWER, ...) in Excel representing a "car model data set" are mapped onto a 3D sphere graph. Multiple 2D and 3D views of the same dataset can often help the understanding of trends and patterns in a multivariate data set. As we become more analytically oriented and begin to think multidimensionally, these 3D techniques will become a necessary tool to deal with massive amounts of complex abstract data.

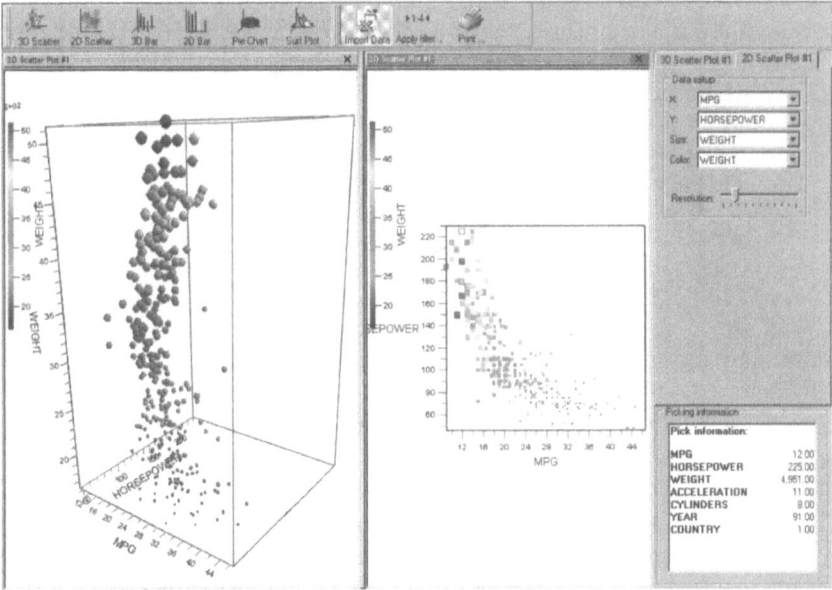

Figure 3.4 The Docking Manager UI component controls the Visualisation Docking Area.

will want to start with the same perspectives that sparked the questions, then apply their own expertise and experience to the analysis by branching off into their own exploration of the data. A useful way to derive insights from your analysis is to create an intuitive map to saved views of the data, and then make this discovery available to colleagues. You will want to create a system that can capture the insights so that they can be shared with the people who can take action on them.

3.3 Visual Data Navigation in a SmartDoc

Users continue to demand better user interfaces that make an application easier to learn, improve productivity, and enable them to gain new insights and understanding of key data metrics. By improving the interface, SmartDoc has managed to shorten the *"time to enlightenment"*. The *"Visual User Interface"* will enable the user to take a more active role in the process of visualising and investigating data. The users interact directly with the on-screen graphics and the data behind the graphics without having to work with traditional GUI controls. The aim is to create a data-centric view, in which the user responds to and interacts with the actual visual presentation of the data on the screen. The on-screen data object is "live" – the object itself includes underlying data structures and properties, not merely a reporting window. The sense of immediacy and speed-of-thought interaction in both 2D and true 3D space (and the correlation between these two) helps users to gain insight. For example, by clicking on a graphical object you are able to *drill down* to more detailed information about that object. Other useful interaction techniques include brushing and hyperlinking. *Brushing* allows the retrieval of more detailed information about parts of the visualisation without changing the visualisation itself. By moving a pointing device over a particular component in the visualisation extra information appears on top of the selected object

3.4 Multiple Views

Most data are not best analysed through the use of a single type of graph. In order to detect complex patterns within the data it is necessary to view it through a number of different visualisation tools, each of which is best suited to highlighting different patterns and features.

If an information visualisation framework is to handle the broadest range of applications, it must achieve a balance between two- and three-dimensional functionality. Some problems are best solved with one or the other; however, many require the services of both technologies. We have expended considerable effort in developing a balanced approach to information visualisation. Where possible, each three-dimensional tool has a two-dimensional counterpart and the object-oriented nature of the technology ensures that most functionality is shared between the two. By using the two together, the power of each is amplified.

The simplest manner in which to employ one or more graphs simultaneously is to view them side-by-side. However, the *context* of each point is lost in the process. One point may be salient in one graph, but may not be identifiable in another. Only through interaction may points like these be located and investigated. The user may select a point that appears interesting in the line chart to see it labelled in the scatter plot. The converse may occur as well. This point may then be deselected from the line chart and the impact upon analysis performed on the data set as a whole visualised in an entirely different graph. Therefore the data, analysis and visualisation "flow" together in a seamless process of discovery. For example, Figure 3.4 demonstrates the correlation between multivariate data visualised as both 3D spheres and 2D scatter plots that identifies how they relate to each other.

Our components consistently implement a Model–View–Controller architecture. This implies a separation between the data and the views and analysis of those data. This simple design enables users to achieve a high degree of interactivity with multiple graphs that visualise the same dynamic data source in multiple 2D or 3D views. The SmartDoc "Docking Manager UI" component allows you to create multiple views on multidimensional data sets. Since each of these view shows the same data, but is otherwise independent of the other views, you can change the setting for each individual view to highlight a different aspect of your data. The white area in Figure 3.4 displaying 2D and 3D scatter plots is called the "Visualisation Docking Area". Any opened visualisations will share the available space in this area. You can rearrange the position of individual visualisations by left clicking on their title bar and dragging the window around. If you have multiple visualisations open at the same time, you can use this feature to get the layout you desire. This feature, in combination with dragging visualisation windows around to rearrange them, allows you to get nearly any non-overlapping screen layout you might desire. We have found that this Visualisation Docking Area is a key feature for multidimensional visualisation.

3.5 Reference Model for SmartDoc

The basis for effectively applying all visualisation techniques and data handling capabilities centres on understanding the concepts and operation of a data flow network (Card *et al.*, 1999), visual mappings and the scene tree. We think of the data navigation process as adjustable mappings from data to visual form to the human perceiver. The diagram of the mappings in Figure 3.5 serves as a simple reference model (like other methods). Raw data, that is, data in some foreign format, is mapped into ordered data tables (Excel spreadsheets). Data is filtered or clustered. Visual mappings transform the data into a scene tree. Finally, the scene tree is rendered into views specified by parameters such as camera position, scaling and rotation. The SmartDoc VUI controls the entire process from data to 3D views. Our VUI uses direct manipulation to interact with "3D Views" (rotation, zoom, translate), "3D Visuals" (pick any graphics objects, data correlation) and "Data Cubes" (filter, cluster etc.). The data is stored either remotely in the server or locally in the client, while the visualisation process (3D Visual Mapping and Rendering) always takes place at the client-side.

In the reference model, visual data navigation takes place at three different stages in the transformation process. For example, data transformation can order or classify data differently. By changing the visual mappings users select different visual forms to view the same data from the data tables. Finally, altering view transformations include simple interactions like zooming or

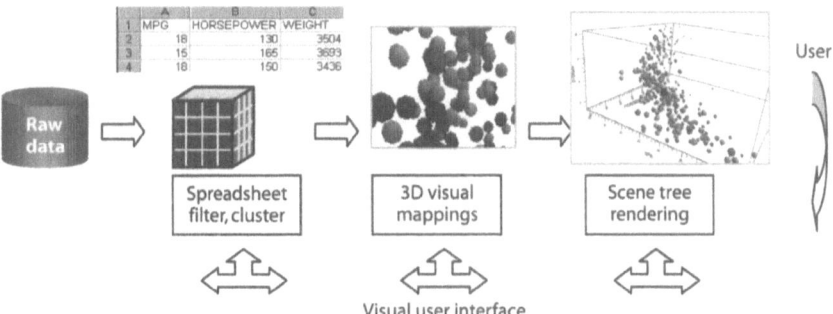

Figure 3.5 The Visualisation Reference model: the visual data navigation in a SmartDoc scenario can be described as the mapping of data to a visual form that supports human interaction in a workspace for 3D visual sense-making.

modifying the viewpoint in a 3D world, and also user interaction such as brushing.

3.6 Building a SmartDoc

In order to embed an instance of the selected SmartDoc component in, for example, Microsoft Word, you use the Insert | Object menu. A list of all components that are installed on your system will be displayed. Selecting "Scatter3D Control" will create an instance of the control and you have an interactive visual data navigation application component embedded in your document. Figure 3.6 shows an example of using the Scatter3D Control.

As Microsoft Excel is the predominant data source in most companies, we have chosen in our project to use an Excel Interface component to import data. This import procedure is done only once; afterwards, the data is stored as part of the document as a 2D array. The Data Import wizard will prompt you to open the Excel document, which contains the multidimensional data you want to visualise. In the next step you enter the data range that you want to analyse. The data is normally embedded in the document, but you can also retrieve the data from a local file or from a central data warehouse.

SmartDoc will provide an image of its contents using a bitmap representation. SmartDoc uses this image for the cached representation of the view when the component is inactive and for printing the document. Finally, if the recipient's

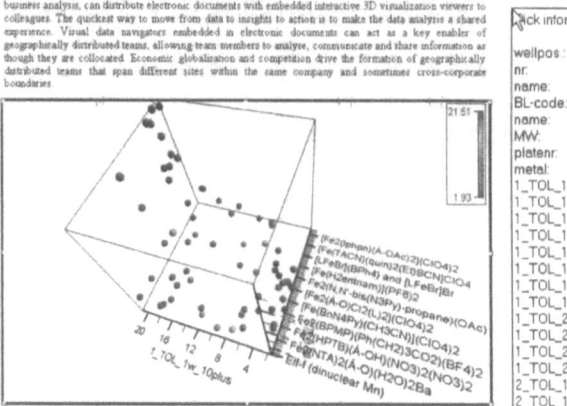

Figure 3.6 he 3D Scatter SmartDoc showing a large multidimensional molecular data set. Observe the annotated axis displaying names of the selected molecules. The box to the right is the result of picking on a sphere.

system does not have the component installed, the user will at least see the image, though he or she will not be able to interact with the data.

When preparing the SmartDoc, the author can set "bookmarks" or snapshots that highlight data views of particular interest to different recipients. Colleagues can use these descriptive bookmarks to quickly locate key information by simply selecting the SmartDoc view they need. The Snapshot Manager remembers and records the status of a data navigation experience. The author has selected suitable data dimensions and display properties, and has filtered data with the slide rangers focusing on the data of interest, finally highlighting the "discovery" from a certain angle (viewing properties). This status "bookmark" can now be saved with the Snapshot Manager. When the document is saved, the snapshots and data will be stored as part of the document. When the next user opens the document, it will start the 3D interactive visualisation process based on the author's bookmarks. The visualisation will revert to exactly the same status as defined by the author. The recipients can then immediately share the author's insight about data.

3.7 SmartViewer

SmartDoc Collaboratories share an "engine" component, responsible for visualisation, interaction and rendering, called the "SmartViewer". This viewer component is a central part of the SmartDoc collaborative process and will become "freeware", allowing the exchange of SmartDoc components between researchers and engineers. The SmartViewer takes care of managing all of the visualisation inside the view, including taking full advantage of and managing the complex interactions with the high-performance graphics layer in OpenGL. The SmartViewer architecture will allow lightweight SmartDoc components to be deployed across the Web. The SmartViewer has a size of about 2 Mbyte, while the SmartDoc components have a small footprint: 50–150 kbyte. Downloading SmartViewer to the client machine enables this. If SmartViewer is not present, it will be automatically installed the first time a SmartDoc is downloaded. All SmartDoc application components share a single installed engine.

3.8 SmartDoc in Practice

We have tested out a "3D Sphere" SmartDoc on several typical multivariate data analysis problems. The following example illustrates a typical generic use – linking between two complementary multivariate data sets as part of the product design process: one experimental and one theoretical.

3.8.1 Discovering Bleaching Agents

An important step in the design of effective laundry detergents is the choice of bleaching agent. This is the key active ingredient responsible for the bleaching

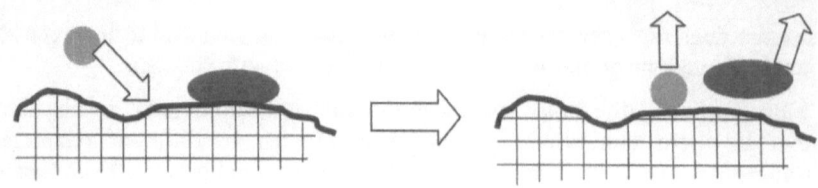

Figure 3.7 The bleaching process.

effect. Bleaching is a molecular process, not a physical one: the beach chemi-
cally interacts with the stain or fabric surface, releasing the stain species
(Figure 3.7).

The overall effectiveness of a possible active ingredient will therefore depend
on three things:

- its molecular descriptors (electronic, electrostatic, structural etc.)
- the in-use conditions: for example the active concentration, the pH, presence of other active species, or the length of time the fabric is exposed to the
  bleach
- the judgement of the consumer who visually assesses the appearance (and
  possibly other parameters)

Our problem then is to link data describing these three areas with the aim of
perhaps identifying novel new active ingredients or the optimum conditions
for effectiveness of existing ones.

In the example here (Figure 3.8), a large database of molecules is screened for
effectiveness in removing dissolving tomato oil stains. This is an initial screen;
the stains are simply fixed in solution – they are not attached to a fabric (any
molecule showing some effectiveness in this first screen would be tested on
actual fabric stains in a similar way).

The molecular descriptors for each candidate are calculated using quantum
mechanical techniques. These descriptors fall into two classes, being about either
the electronic properties of the molecule (e.g. the chemical bond properties) or
its structural properties (its shape); both are likely to matter in bleaching. We
note in passing that for a large number of candidate molecules this data set alone
is very large and is ideally suited for exploration using the navigator, should the
user be interested in finding correlations among molecular structures alone.

Collecting data on the in-use conditions and consumer perception is classically
much more time-consuming. To speed things up a high throughput methodology is adopted using a 96 well plate (Figure 3.9). This is an array of small cells,
each of which can contain a few millilitres of a TOL sample at a given concentration, pH etc. At time = 0 a candidate bleaching molecule is introduced into each
cell and any colour change monitored as a function of time. The bigger the
colour change, the more effective the molecule as a bleaching agent for the
given stain. This is the measure we use as a substitute for consumer perception.
Such an experiment can be highly automated and repeated continuously.

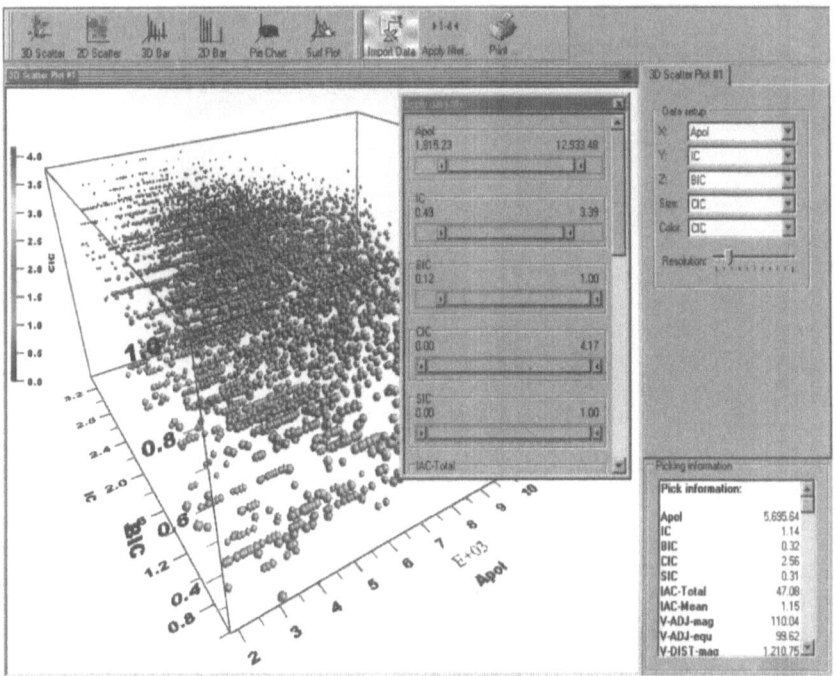

Figure 3.8 The Data Discriminator "Filter" and molecular descriptors. The filter component is used to reduce the data displayed in the SmartDoc. The filter menu will open a window with range sliders, one for each column in your data file. This allows you to adjust the minimum and maximum data values that are displayed for each column. A data value (a glyph representing a row in the spreadsheet) will disappear from your visualisation if any one of the values in this row is blanked out by the slider associated with it. The data contains electronic and structural descriptors of a set of candidate bleach agent molecules.

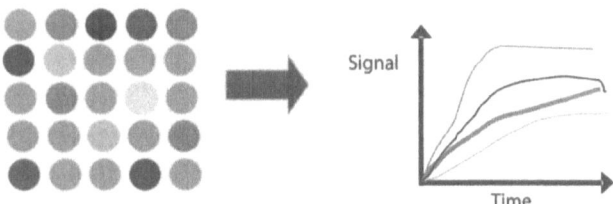

Figure 3.9 High throughput screening. A time series signal is recorded from each cell in the 96 well plate. In this case the cells are lit from below and the signal is proportional to the amount of light transmitted through the sample, so the higher the signal the more effective is the candidate bleach agent in dissolving the stain.

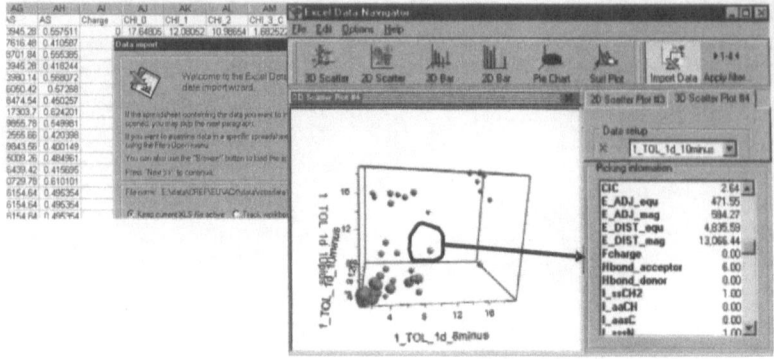

Figure 3.10 Initial data import. Each glyph relates to the contents of one particular cell.

The data is collected within an Excel spreadsheet and merged with the molecular properties data. We then have a multivariate time series data set containing molecular, in-use and bleach effectiveness information to investigate.

Loading the data into the navigator using the import wizard (Figure 3.10) gives us an initial view, which, for a 3D scatter plot, uses the first three columns of data to establish the coordinate axes. The first three columns in this case are appearance measures under three different in-use conditions. Drill-down allows us to find every property associated with the contents of a particular cell in the 96 well plate.

What we are initially viewing is part of the space of "in-use" conditions. Our first aim is to try to identify the type of molecule that might be a good active ingredient, so we first change interactively to a view where the axes are defined using three of the molecular descriptors. Any point within the space of these axes then represents a particular molecular configuration, whether or not that configuration exists. We can then map the bleaching effectiveness under a given set of conditions onto this space of molecules and instantly see the performance of each active molecule. Mapping to different bleach times, concentrations and other experimental conditions enables us to quickly confirm the general rules for bleaching effectiveness; for example:

Longer time   → Better bleaching
Higher concentration   → Better bleaching

The three molecular properties we happened to choose to define the axes here clearly show no correlation (Figure 3.11). However, we can map a further molecular property into the visualisation by using the glyph size variable. The application enables us to quickly scan through all the calculated molecular properties. Something interesting happens when we map the ability of the molecule to accept or donate electrons from or to its environment to the glyph size (Figure 3.12).

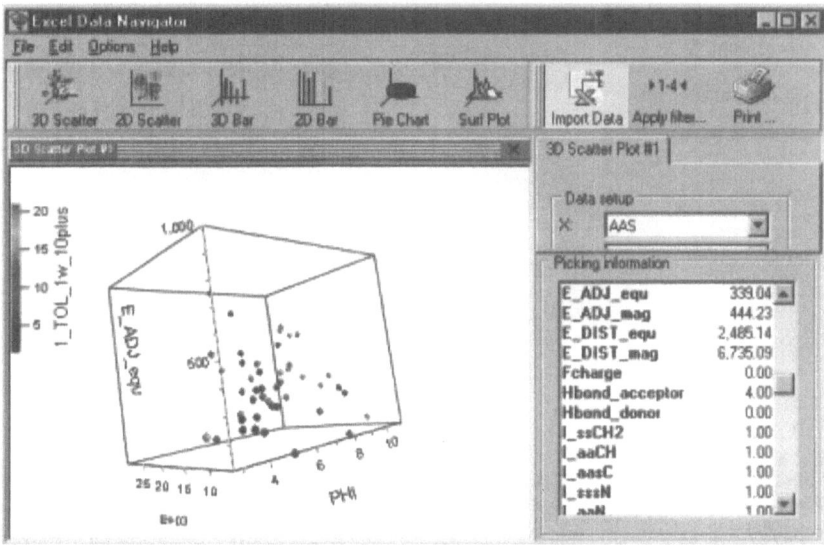

Figure 3.11 After swapping the axes so that they now represent three of the molecular descriptors we can use the glyph colour to represent the amount of stain removed after a certain in-use condition (in this case the amount of stain removed after 1 week at pH 10 without any additional accelerator molecule). In these conditions the successful candidate molecules are highlighted yellow–red. There is no real clustering either of colour or position, indicating that the molecular descriptors used for the axes do not correlate well with the molecular bleaching capability.

Clearly for most cases there is a strong correlation between these electron acceptance/donation properties and the effectiveness of bleaching. This leads us to attempt to produce a more effective visual clustering by making the electron acceptor property one of the spatial axes. Further sampling of the molecular descriptors enables us to cluster the effective candidate molecules into a much smaller region of the configuration space (Figures 3.13 and 3.14).

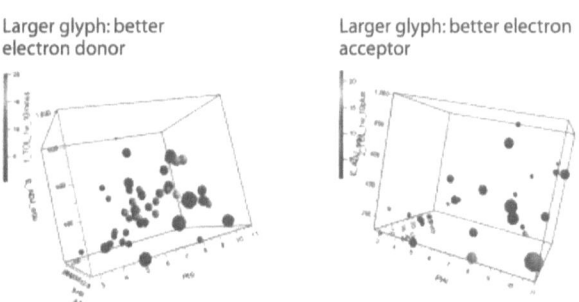

Figure 3.12 The vast bulk of effective bleaching agents have similar – rather limited – electron donator/acceptor properties.

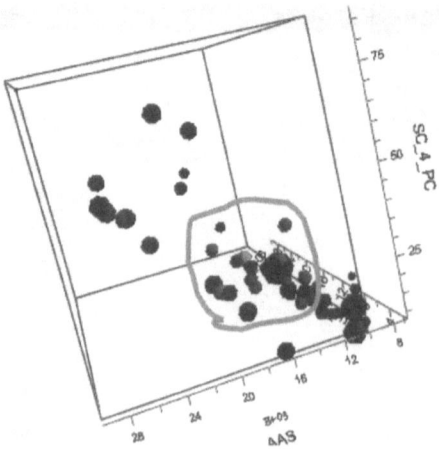

Figure 3.13 Exploring the molecular descriptor space further allows us to pin down the successful candidate molecules to a much more limited region of the configuration space. Colour is again a measure of the effectiveness of the molecule at bleach removal.

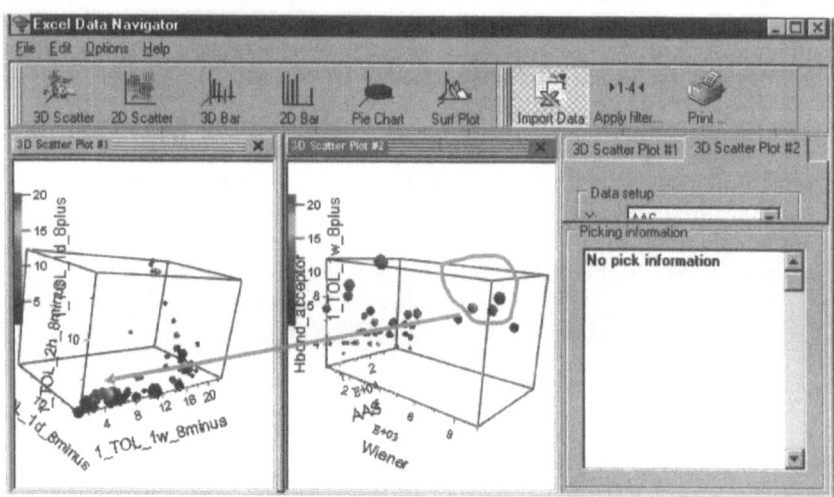

Figure 3.14 Two coupled scatter plots which illustrate the effect of the additional additive. In both the glyphs are coloured by the signal strength with additives. The first plot (left) is of the space of signal strengths without additives. If the additive made no significant difference we would expect the colours to vary smoothly from blue at the origin to red at the extremes – but there are clearly two clusters. Further, by picking we can see how these clusters map into the molecular descriptor space.

3.9 A SmartDoc Scenario

Step 1: Create the spreadsheet data. Transform raw data into data tables.

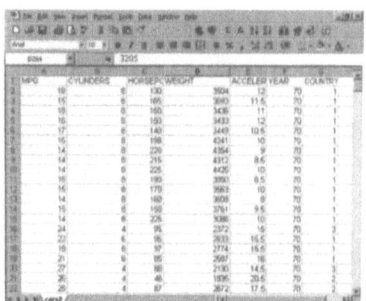

Step 2: Select and download visual data navigator application components ("Collaboratories") from a central portal component repository. Download the Smart-Viewer if not already installed.

Step 3: Embed one or several selected Collaboratories in, for example, a Word document by using the Insert | Object menu. A list of all components that are installed on your system will be displayed. Select "Scatter3D Control" and this will create an instance of the control and you have an interactive visual data navigation application component embedded in your document.

Step 4: Open the Collaboratory and visualise interactively the data inside the document. View, zoom, select attributes, pick data and isolate data within certain ranges by filters until the author has discovered pattern or trend in his data. *Save snapshots at each level of discovery.*

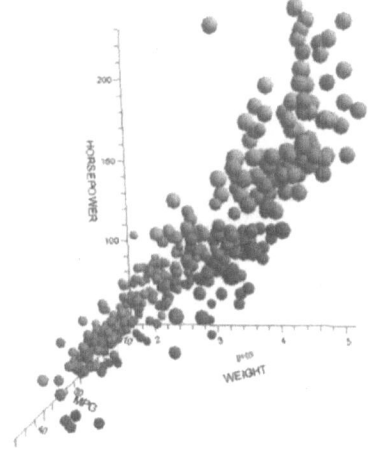

Step 5: The author is now ready to create the final SmartDoc document: text (content), final image (selected view of data), status file for the snapshot (selected data and visualisation attributes are saved), visualisation components (pointers), and data (embedded as an "array"). The author sets snapshots that highlight data views of particular interest to different recipients. Colleagues and clients can use these descriptive bookmarks to quickly locate key information by simply selecting the report view they need. Recipients also have the flexibility to create their own bookmarks for at-a-glance reference and comparison.

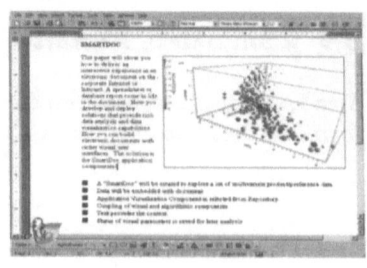

Step 6: After the author completes the report, he or she sends it to both internal and external constituents who have access to the SmartViewer. The document is also posted to the organisation's Web site. Now internal managers and marketing researchers etc. can interact with the document on the organisation's Intranet while external clients can work with the document via the Internet.

Step 7: Recipients retrieve the SmartDoc document from the Web site. The visualisation components and SmartViewer are downloaded to the client machine (if not already available).

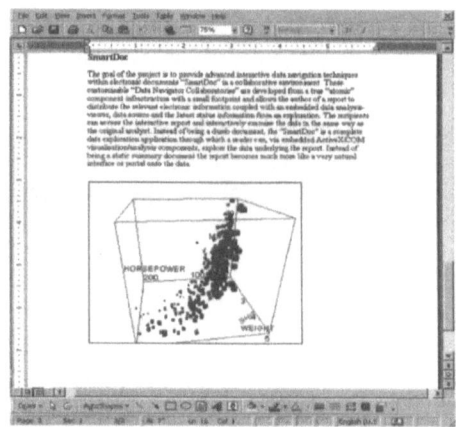

Step 8: Recipients can now dig into results and get more value by creating new meaning and understanding in the results:

- Start the analysis from a snapshot
- View the analysis behind the report

- Interact with the results and digest
- Change visualisation parameters
- Discover new meanings of data, trends and correlation
- Recipients are more active than with just text reports
- Promote collaborate research
- Generate a new SmartDoc report

3.10 Conclusion

The prospects for SmartDoc and Collaboratories look promising. In particular, the Web has created a model for disseminating information. The traditional "thin" client–server model, however, works poorly for data interactivity required in a data navigation scenario. The standard Web interface encourages data navigation and its visual user interfaces to focus on the lowest common denominator, which would require high bandwidth. The emerging visualisation component architecture based on "atomic" components with a small footprint could, however, move data navigation techniques from research to products.

The SmartDoc EC-funded research project has demonstrated how to deliver an interactive experience in an electronic document on the Web based on embedded interactive visual data navigation components: "Collaboratories". Upon receipt of a SmartDoc, the recipients can immediately share the author's insight and zoom in on specific results that affect their activities. Clients and colleagues can rearrange multivariate data to uncover patterns and trends that may go unnoticed in lengthy paper reports. The distributed architecture is based on "Application Component Sharing", providing real-time data interactivity, reducing the load on the network and with zero-administration client deployment.

Also important is the aim to promote the use of a component-based approach to the development and engineering of software systems, applications and services (Figure 3.2). Customisable and scalable high-level "application" and "functional" components were designed and developed from low-level "atomic" components. Our Collaboratories are based on Advanced Visual Systems' OpenViz, a low-level visualisation component framework. Atomic components from several other sources, including data interactors, data filters, analysis and data access were also integrated. We believe that using lower-level atomic components for developing application components would provide better scalability and more customisable visual data navigator components. Atomic COTS components from different vendors (or developed when necessary) were used in assembling functional and application components.

We have shown the possibility of deploying 3D visualisation in electronic documents. Based on our experience, we have drawn some tentative conclusions regarding 2D versus 3D data visualisation. We can conclude that 2D data visualisation methods are more easily accessible to the user. The 3D data visualisation allow the user to combine more information into a single scene, but these methods are not yet accepted as instruments for decision-making among the business community.

Another overall goal of SmartDoc is to make people more effective in their information or communication tasks by reducing learning times, speeding performance, lowering error rates, facilitating retention and increasing subjective satisfaction. We believe that customisable and scalable Visual User Interface (VUI) components in collaborative work can increase effectiveness for users who range from novices to experts and who are in diverse cultures with varying educational backgrounds.

SmartDoc will be tested and validated in European global industries with geographically distributed remote users, linking people and their desktops (Collaboratories) in a worldwide "Virtual Data Environment". Customised medical imaging application Collaboratories will also be tested in European healthcare. We believe that SmartDoc can contribute to a new wave of accepting advanced real-time data visualisation and navigation in a collaborative environment, with the focus being on the people using the system.

SmartDoc's features include:

- Dynamic multidimensional graphics that users can interact with in real time.
- A rich set of component resources with granular control of details.
- Platform/rendering library independence through the SmartViewer plug-in.
- Visual User Interface – tight integration between data and visualisation objects.
- Optimised rendering based on scene tree hierarchical graphics structure.
- Binning, filter, crop, sort and aggregation to achieve a good interactive response.
- Drag, rotate, zoom and pick empower users to explore data.
- Data reduction integrated into visual data navigation.
- Full integration with the data warehouse or spreadsheet.
- Presentation graphics with axes, legends, annotation and high-resolution hard copy.

Acknowledgements

This paper was supported by the European Community in the ESPRIT Project CONTENTS (EP 29732) and Advanced Visual Systems.

Reference

Card, B. *et al.* (1999) [Is this the one? Card, S. K., Mackinlay, J. D. and Shneiderman, B. (eds.) (1999) *Readings in Information Visualization: Using Vision to Think*. San Francisco: Morgan.]

4

The Future of Convergent Computer and Telecommunications Technology

Peter S. Excell

Abstract

This chapter reviews, and raises questions about, the way in which personal computer and mobile communications technologies will be used in the near/medium-term future. The evolutionary paths of the relevant technologies are reviewed and it is argued that they are converging rapidly. Inasmuch as mobile communications have a clearer forward plan and a larger sales base, they appear set to become a dominant paradigm. However, the lack of vision regarding the way in which they might be used, and the slow rate of development of appropriate content suggests that the industry needs to consider the wider implications and to think very radically and creatively about future applications.

4.1 Introduction: Where Have We Come From?

Electronic computers were first developed just over 50 years ago. They were initially seen as essentially mathematical calculating engines, and hence were located at first within university mathematics departments. Nowadays this view is seen as being more or less misguided, but it persisted for 20 or more years. And yet the seeds of an alternative view were present in the earliest days, although not generally recognised as such. The alternative view is of the computer as a communications device, converting information from one form into another: this was the function of the now-famous Bletchley Park "Colossus" computers, but the secrecy surrounding them seems to have prevented many valuable lessons migrating into the civilian domain. In the late 1960s and early 1970s, work started on the development of computer networking and the forerunner of the Internet, and yet initially this was mainly seen as a way of logging on to a remote computer, usually one that was more powerful than the best available at the user's home site.

Starting a little before the development of networking, and driven by a communications technology need to improve signal-to-noise ratio, telephone networks started to adopt digital (pulse code modulation) systems of transmission, which meant that the exchanges could be converted to a digital form with stored program control, making them a form of computer, although generally considered to be a very "special purpose" device.

Slowly, communications uses of mainstream computers started to be developed, always initiated as "serious" applications: e.g. email for government messages and the Web for nuclear physics images. Personal computers, again, started with the expectation that applications would be "mathematical"; ideas for uses in the home were relatively uninspired and the business almost died (to quote a respected technological forecaster, writing in 1988: "By the mid-1980s the home computer... was a spectacular failure..."; Basalla, 1988).

Then came a great transformation. With the end of the Cold War, funding for "serious" uses declined and computer and software manufacturers realised that they had to please a mass commercial market if they were to stay solvent. "Frivolous" uses became acceptable; profit margins became extremely thin, but turnover grew massively. Despite fears that the relative decline in funding for "serious" applications would strangle the business, the massive growth in consumer applications threw up new challenges and began to suggest deeper implications for a "fully networked society".

Communications, meanwhile, had been a sleeping giant. Digital transmission had been introduced, but mainly to improve signal quality, rather than for reasons of integration with computer networks. High-bandwidth backbone links had been developed, building on the revolutionary bandwidth increases possible with optical fibres. The huge capacity generated led to the possibility of exploitation of small and unpredictable opportunities of idle time to transmit packetised messages: these had greater robustness and lower cost than circuit-switched approaches, but they crucially depended on the existence of a computer, of some form, at each end of the link, in order to packetise and reassemble the transmitted message. Fixed wire line communications to mass-market consumers are currently still mired in the problems of exploitation of the traditional copper local loop, but the potential of the technology, as enhanced by computers at its nodes, has been demonstrated by basic Internet access, streaming audio, software downloading, Internet telephony and Web cameras, to mention but a few applications.

Then came a more major communications revolution. Perhaps because it started in Europe it has not yet been given the public relations profile, as a concept, that it merits: the revolution was GSM (http://www.gsmworld.com/technology/data_services.html). Initially named Groupe Speciale Mobile, it was later rebranded in an anglicised form as the Global System for Mobility, presumably to improve its international acceptability. It is, however, hard to overestimate the significance of the work of this rather anonymous committee, which, starting from a desire to improve the performance of mobile phones and overcome the deficiencies of the first generation analogue systems, developed a scheme that looked breathtakingly complex in its time and yet, because of the availability of cheap "computers" (in fact, DSPs) it could be implemented at modest cost and deployed widely in the mass market. Much more than that, however, it set an entirely new pattern that points the way to an expansion of capacity while building on the same principles. As a result, the mobile telephone paradigm is set to overtake the wired approach for many purposes. Further, the functionality of mobile communications devices, in the near future, is likely to be such that they will subsume the personal computer business. It is the epitome of the concept of convergence between computer and communications systems and it is a technology that is truly changing the world, in both developed and developing countries.

The mobile telephone has now become an established part of everyday life, not only in developed countries, but also across a wide area of the developing world. This is clearly illustrated by predictions that the worldwide annual sales of new mobile phones will reach 740 billion by 2004 (http://www.it.fairfax.com.au/communications/20000926/A13234-2000Sep22.html), and many commentators expect this prediction to be fulfilled before that date.

4.2 Where Are We Going?

The mobile phone of today has more processing power than the personal computers of only a short time ago. It has a degree of Internet ability (rather unsatisfactory at present, because of a misguided attempt to tailor it to the limited human–computer interaction abilities of the device): all it needs to essentially supplant the personal computer is storage and better human–computer interfaces (HCI). The solution of these problems is not "rocket science". Miniature hard disks are already available: input techniques such as voice recognition, handwriting recognition and folding or miniaturised keyboards are well known, and ideas for larger video displays within mobile phones, or the use of separate retinal projection devices, are established in the research laboratory.

Filled with confidence by its success, the mobile phone industry has generated a raft of ambitious roll-out plans for improved services, based on the principles established by the GSM system, which represents the 2nd generation ('2G"), analogue cellular mobiles being the first generation. These improvements are collectively known as "2.5G" as they represent a half-way house on the way to the Third Generation ('3G", also known as the Universal Mobile Telecommunications Service (UMTS); http://www.umts-forum.org/). Thus we have already seen SMS (Short Messaging Service) and WAP (Wireless Applications Protocol), representing software-based extensions of functionality, plus HSCSD (High-Speed Circuit-Switched Data) as an example of a partially hardware-based extension, permitting some acceleration of data rates over the basic GSM rate. HSCSD has not proved very influential, as it can only be used through the metered-call mechanism, but the next hardware extension, GPRS (General Packet Radio System) will be more so, because it will permit unmetered access, as well as a modest increase in data rate. Following that, the proposed next step is EDGE (Enhanced Digital GSM Environment), which would permit a substantial data-rate increase, but as it will require substantial hardware upgrades at the base stations many believe it will not be deployed and that investment will switch to 3G systems, which have an additional spectrum allocation, high data rate and an Internet Protocol basis for packet-switched data. Such systems have the ability to deliver videophone and mobile multimedia functionality, at least when a base station is within close range, although various groups are working on ways to increase this range. The target date for 3G roll-out has been around 2003 for some time, but this may move forward as a result of a perceived need, in countries which did not fully embrace 2G, to leapfrog 2.5G and invest in a new system with more stability. Stability, however, may be illusory, as several research organisations are already working on an all-IP-based Fourth Generation (4G), to be deployed before the end of the decade (http://www.mobilevce.co.uk/core2/core2frame.htm).

With massive internal processing power and IP connectivity, projected forms of third and fourth generation mobile communications equipment are potentially strongly competitive with the personal computer paradigm. All they lack, at least in the basic "mobile telephone handset" format, are storage and competitive HCI. Storage can easily be provided by using devices such as "microdrives" (very small hard disks). Input with some kind of folding miniature keyboard could be arranged, but it has to be admitted that the QWERTY keyboard is far from ideal as an input device and methods such as a mouse-like device, handwriting recognition, speech recognition and predictive text input from a 10-key keypad may well be more suitable for mobile devices. Output could be, for example, to an enlarged screen such as on a personal digital assistant, or to a retinal projection device. However, doubts remain about the communication efficacy of such small-screen devices and it may be that a relatively large unfurlable flat screen, as has been suggested in some prototypes, may be found preferable. Resolution of this issue is urgent, as the public has come to expect large-screen interaction with the Internet: the WAP approach to small-screen interaction has been found unsatisfactory, but incremental extensions to this very limited interface seem unlikely to be acceptable as a settled paradigm until they reach near-equivalence to the large-screen format. Retinal projection certainly seems to offer the most compact approach to this, provided the eye can feel comfortable browsing the content of the equivalent of a large screen of information: the exemplar of the video camera viewfinder (of traditional type with a small screen and a lens) seems to indicate that this will be possible. The essential point is that the gap between these advanced mobile communication devices and the concept of the personal computer is very small and is not difficult to bridge by modest extensions of proven technology.

In contrast, the personal computer industry currently appears to lack a clear vision for its future development, possibly as a result of the pressure of competition on its profit margins, and this suggests the opportunity for future generations of "mobile phones" to take over many of its functions (The term "mobile phone" is seen to be becoming rather restrictive and it is desirable that a new name be found to indicate its processing power and intelligence. However, it will be used here for convenience). This vision is not simply driven by a "geeky" push for more and more advanced technology, although the importance of enthusiasts in exploring future options should not be underestimated. Probably the most exciting implication of the convergence of the "mobile phone" and "personal computer" paradigms is the potential to "digitally enfranchise" a far wider spectrum of global society. If we compare the social groups currently predominantly using these two types of technology, it is reasonable to say that personal computer usage (at least, outside the workplace) is male-dominated and largely confined to relatively developed and fast-developing countries. In contrast, the mobile phone has a far higher female user base and it also has a far higher usage in the less developed parts of the world. This means that, even from a raw commercial viewpoint, the client base is far larger for the "mobile phone" paradigm; hence the economies of scale are far greater and can be invoked to defray the research and development costs of future novel systems more readily than for extensions of the "personal computer" paradigm, especially since the skilled users of the latter have come to expect enhancements to be very keenly priced.

With due caution, it is worth exploring why female and Third World uptake of the mobile phone are proportionately so much higher than for the personal computer. Firstly, it has to be admitted that they are currently effectively different devices, even though an understanding of technical trends indicates that they will converge in the near future. The main attraction of the mobile phone is clearly very convenient pervasive communication. It is a symbol of association with high technology that is not overly costly and which can be carried by the user, making it more secure than the PC. It is not dependent on unreliable cable and power systems. The strengths and weaknesses of the personal computer need to be broken down, dependent on different usage modes. Email is clearly a personal communications mode, specifically non-real-time, but this mode has now been widely adopted by mobile phone users in the form of SMS, and extension to longer messages will soon be delivered by 2.5G systems. Addition of video functionality to messaging has been piloted by "Webcam"-type peripherals for personal computers, not necessarily used in real time, despite the implication of the name. These have not been widely taken up, due to difficulties in the user interface, but it appears that third generation mobile phones will overcome this problem. Internet browsing, including e-commerce, to procure goods and services found, is clearly well established on PCs, but its migration to mobile devices is well in hand, although the scale of the traffic is likely to be much smaller than for personal communication. Gaming is an important area, although this is likely to be mainly of interest to relatively affluent males. Relatively passive television-like functionality (e.g. streaming video and non-real-time video, although the former is currently constrained by bandwidth problems) is likely to be a larger market than games, although there could be a degree of convergence between the two modes, since both are forms of video entertainment and could be made to be very similar if degrees of interactivity could be made variable and optional. Graphical/multimedia artifacts/content generation are clearly major usage areas for PCs that seem specific to the personal computer model, and this may remain the case, since there is likely to be little demand to undertake them in a mobile format. However, again, the economies of scale may mean that the mobile device becomes the lower-cost technology and it is possible that it could become the norm for non-mobile use as well when this is appropriate: the folding keyboards that can be attached to PDAs are a simple example of such a usage scenario.

4.3 Future Concepts

The clear logic of computer–communications convergence is that a high-end future system will have processing and storage capabilities similar to a good current PC and third generation communications ability of several hundred kbps. Adding a convenient and not-too-conspicuous human–computer interface means that it will have to be distributed around the body, in other words it will become the same as the "wearable computer" with a wideband radio modem added (see, for example Ditlea, 2000; http://www.digiman.org/html/hmd.html; http://wex.www.media.mit.edu/projects/wearables/; http://www.xybernaut.com/). Such devices, with narrowband or non-existent radio links, are already in limited use, but mainly as an aid to certain types of manual employment (Ditlea, 2000).

If this kind of technology is to become acceptable and used, at least in employment, by a wide spectrum of workers, and by the general public outside the working environment, more radical ideas need to be considered for the way in which it would be used and, in particular, the advantages it may confer that would justify its purchase and use. Firstly, at the practical level, it is obvious that the cabling that it requires is objectionable and this would be likely to be replaced by Bluetooth links (http://www.ericsson.com/infocenter/news/Bluetooth_headset.html; http://www.bluetooth.com/developer/specification/core_10_b.pdf) between a CPU, a possible separate storage unit, a combined retinal projection unit, earpiece and microphone, a wrist-mounted keypad/mouse unit and a radio transceiver which could be worn on the belt, but could equally well be integrated into a shoe from which electric power could be drawn (http://www.theelectricshoeco.com/news.html). If, however, links to satellite mobile radio systems were required, the transceiver would have to be near the head. The whole would form a Personal Area Network (PAN).

How would (will) such a system be used? Video phone communications on the move? Certainly, but not all of the time. Email, with optional added pictures or non-real-time video? Yes, this is likely to be a major source of traffic. Web browsing? Certainly, but not much of the time. Provision of information relevant to one's employment? Very possibly, but not outside working hours. Immersive gaming? Possibly, but not for the majority and not in working hours. Is this all?

4.4 Conclusion: A Possible Vision of the Future

To quote the Sun™ strap line: "the network is the computer™". The bandwidth of third and fourth generation mobile systems will allow relatively good integration with a global network of computers and computer-based communications devices. Each individual node will have a substantial degree of intelligence, but the knowledge and intelligence of the whole network will be very much greater and will grow rapidly. Thus the person wearing a wearable computer with good HCI and wideband radio communications will be relatively tightly integrated with the global distributed intelligent information system. The knowledge contained in the system is already much greater than that carried by any single human being. Its mathematical processing power has been vastly greater for many decades. Its "intelligence" is questionable, but is improving all the time and is predicted to overtake some aspects of human intelligence within a few decades. The logic of the situation, therefore, is that the human user will become something of an appendage attached to the global intelligent network.

The standard caricatures of this kind of scenario depict a frightening and dehumanising experience, but should it be so? Consider the case of a user with a mental handicap: in principle, a computer could be a valuable assistant that could advise a person with a modest handicap what they should do in particular circumstances and could remind them of facts forgotten due to short-term memory problems etc. With additional input/output devices, in particular video cameras, the potential must surely be present for such integration with large computing power to greatly enrich the life of a person with such a handicap.

But should such aids be limited to the handicapped? If the network has the potential to be more intelligent and knowledgeable than even the wisest human being, surely it has the potential to improve the life of any person integrated with it? Of course, this raises deep questions about the nature of intelligence and the value and purpose of human life. This is a good thing: just as computers and automation have rendered it unnecessary for human beings to spend their lives doing many monotonous jobs, the wearable computer/communications system has the potential to free the human being from the great majority of tedious and non-creative activity. The human would be freed to concentrate on faculties that cannot be replicated in a computer. In fact, the human would become akin to a special-purpose "peripheral", with unique abilities, attached to the computer system. This concept is being explored at Bradford University in a pilot project entitled HAPPI (Human as Peripheral Paradigm Initiative).

The potential exists for this kind of vision to be realisable within about two decades, if not sooner, and it behoves researchers and relevant sections of industry to explore such possibilities and consider alternative visions, of which this is but one, so that preparations can be made, especially in development of the relevant software. The potential of the vision being presented is challenging, frightening and exciting. It seems likely to change human life more than any other technology. Even those who object to it will be unlikely to be able to prevent its evolution.

Acknowledgement

I would like to thank my visionary colleagues Rae Earnshaw and John Baruch for many stimulating discussions and for helping to create the environment that enables the Bradford University School of Informatics to have a strong focus on the potentials of the future. The Bradford-based Technological Fore-casting specialist, Brian Twiss, has also been a great source of inspiration.

References

Basalla, G. (1988) *The Evolution of Technology*. Cambridge: Cambridge University Press.
Ditlea, S. (2000) The PC goes ready-to-wear. *IEEE Spectrum*, 37(10), 34–39.

5

Towards Autonomous Characters for Interactive Media

Daniel Ballin, Ruth Aylett and Carlos Delgado

Abstract

This chapter describes a project to develop virtual environment-based Teletubby characters, as in the children's television programme. The differences between the media are considered and the requirements for character development. The multi-level architecture adopted, BALSA (Behavioural Architecture for Life-like Synthetic Agents), is described in detail and future developments are proposed.

5.1 Introduction

People get very attached to popular television and movie characters; the number of people that flock solely to see Mickey Mouse™ every year reflects this. Some adults have been known to go as far as to disbelieve the fact that Mickey Mouse is just a cartoon. Children often dress up playing heroes and villains together, enacting scenes from films and comics. But the child's inclination to retreat from reality through the diversion of fantasy only lasts for short spells. This is not usually because of any lack of imagination, but typically because they are constrained by the environment and the way they interact with it. The illusion of being Superman or Superwoman is fragile and cannot be sustained, as the boys and girls playing the game realise they are just in a typical back street in suburbia. To be Batman™ or Luke Skywalker you want a realistic environment, having total freedom to do as you want, and inhabit it with suitable characters who are believable and act and react appropriately to your and others' actions.

Virtual Reality (VR) or Virtual Environments (VEs) are a user interface technology that allows humans to visualise and interact with three-dimensional synthetic environments in real time through their sensory channels. "VR is a new medium, as films, and television once were" (Pausch *et al.*, 1996) "by offering levels of interaction and stimulation that conventional media such as television or cinema cannot produce" (Dysart, 1994). With this premise in mind, using high-fidelity VR we might be able to sustain the child's illusion of being somewhere where they might be able "*to become*" one of their fictional characters. This sets the backdrop for a project to produce an immersive virtual world, authentic in look and behaviour with an existing children's television programme. One of our key objectives was to create lifelike characters that populated the world and would interact with one another in a suitable fashion. This chapter introduces the Virtual Teletubbies, and then gives a detailed treatment of BALSA, the autonomous character architecture that we developed and implemented.

5.1.1 Intelligent Virtual Agents

The autonomous characters developed in this project can be described as intelligent virtual agents (IVAs). We mean by this embodied entities each of which autonomously senses its environment – through virtual sensors – uses the sense data to "decide what to do", and then carries out the activity using its virtual body.

These agents are not, in the correct sense of the word, avatars. An *avatar* is, strictly speaking, the representation of the human user in a virtual environment, and is entirely driven by the user without performing any autonomous interaction of its own. In particular, an avatar has no virtual sensors or internal processing of sense data.

An IVA is also very different from the *virtual actors* used in the film industry – for example in the films *Titanic* and *Gladiator*. Virtual actors are developed by transforming a standard humanoid model according to the proportions of a real human actor and texturing accordingly. Motion capture with the real actor is used to animate the virtual actor and its overall movement is entirely scripted, as is appropriate in the non-interactive context of a film.

In a virtual environment, however, there is no fixed duration or linear plot and the user is free to wander and interact at will. Under these conditions, scripting the activity of other characters soon undermines the user's feeling of presence and involvement – where real-time interaction is the norm, characters must also derive their activity and motion from real-time interaction, sometimes known as "self-animation" to distinguish it from standard pre-scripted animation.

5.2 Virtual Teletubbies

5.2.1 The Genuine Teletubbies

The domain chosen was that of the children's TV characters the Teletubbies. The award winning television show of the same name is produced by and ©

Figure 5.1 The Teletubbies.

Ragdoll Productions (UK) Ltd. The multi-million pound series has been sold to 102 broadcasters in 111 countries, including the USA, and has been translated into 41 languages (http://www.ragdoll.co.uk/teletubbies/progr_ teletubbiescoverage.html). There are four characters: the Teletubbies shown in Figure 5.1 (from left to right) are called Tinky Winky, Dipsy, Laa-Laa, and Po.

As agents, Teletubbies are both behaviourally and graphically tractable: they have a limited repertoire of identifiable behaviours, rounded two-colour bodies, simple limbs, single facial expressions, and limited speech capabilities. These characteristics make the Teletubbies suitable subjects to model, and the popularity of the real characters meant that most young children could identify with the virtual characters.

5.2.2 Inhabited Television

Early ground-breaking experiments in *inhabited TV*, where audiences can participate *en masse* in sometimes live TV shows within collaborative virtual shared worlds have been performed (Walker, 1997). In all these experiments the concept of traditional broadcasting was extended, and audiences (sometimes online) could influence the course of the television programme through social interaction. However, one problem it suffered was a lack of engagement between performers and the interacting audience (Benford *et al.*, 1998). The Virtual Teletubbies project does not attempt to mix traditional broadcasting with virtual environments; instead, it tries to immerse users as one of the starring characters in Teletubbyland. As we have already noted, however, a major difference is that a television show is necessarily linear, with a beginning, middle and end, following a story. A VE is much more, with the user's avatar having the freedom to wander and interact. Populating a VE with autonomous Teletubby agents makes the system even more unconstrained. The goal of the project is to faithfully reproduce its spirit and "look and feel", and yet offer a level of interaction and stimulation that the programme cannot.

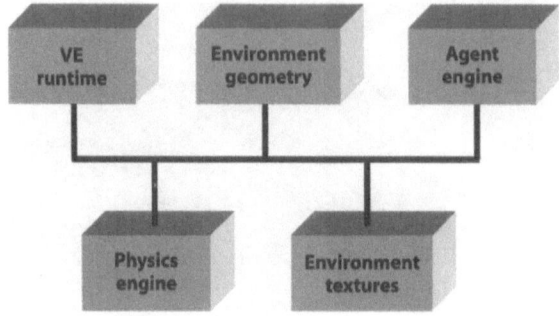

Figure 5.2 Overall system architecture.

5.3 Overall System Architecture

The overall system (Figure 5.2) includes a run-time VE generator that brings the system together. The geometry of the environment includes Teletubbyland and all objects and agents that exist in the world. Objects also carry properties such as colour, shading and behaviours. Most of the external facets within the environment had an associated texture rendered by the run-time engine. Physics in many respects is the essence of reality, and for this reason Teletubbyland has its own very simple physics engine. The rest of this chapter briefly overviews Virtual Teletubbyland, modelling the Virtual Teletubbies. It then shows how we extended a robot agent control architecture to Virtual Teletubbies.

5.3.1 The Synthetic Teletubbyland

The environment in which the virtual agents are situated is bounded and park-like, with the exception of a Dome living area (Figure 5.3). It involves a small

Figure 5.3 A user explores Teletubbyland. The Dome is cut away to show the internal structure.

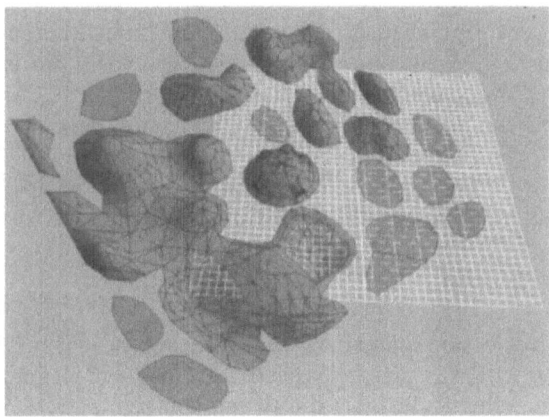

Figure 5.4 Formation of the landscape of Teletubbyland.

number of other devices – for example the Noo-noo (a vacuum cleaner creature on wheels that goes round sucking messes up), a windmill and a motor scooter (ridden by Po). The Noo-noo can be thought of as a fifth Teletubby with behaviours required to notice "mess" and clear it up. While the windmill and the toaster (for making Teletubby toast) have limited capabilities, they actually use the same agent-engine as the Teletubbies, but each has one extremely primitive behaviour mapped on to a trivial actuator. This highlights the strength and pliability of the agent-engine, in that behaviours for simple dynamic objects in the virtual world can be fashioned quickly. The landscape was formed around the Dome by manually creating a series of hills and valleys of random size and shape. A shot of the formation of the land can be seen in Figure 5.4.

5.3.2 Modelling the Virtual Teletubbies

Each of the Teletubbies is built up of a hierarchy of limbs acting as an invisible skeleton. This made it possible to control the motion of the agent using the kinematics of the jointed structure, in the way that robot arms are controlled. However, inverse kinematics has a relatively high computational cost (partly due to the need to carry out matrix inversion), and it is important to consider whether this cost is worth the return. In the case of photo-realistic virtual agents such as semi-autonomous virtual actors and virtual humans (Boulic *et al.*, 1995; Thalmann and Thalmann, 1998), observers are relatively unforgiving if the agent gait is not consistent. However, for Teletubbies, the natural behaviour of the TV characters is to waddle when moving. Therefore it was deemed overkill to incorporate a sophisticated kinetic model, and a motor control layer is built into the control architecture described later.

Developing *believable* characters was important: believability seems to be the summation of two key components in synthetic characters: visual realism and behaviour. We followed the concept of "rendering behaviour, then pixels" (Ventrella, 1998), letting the agents' demeanour and interactions make the user feel that it was real. The majority of computational time was thus dedicated to the agent behaviour architecture and less time was spent processing detailed shadows and aesthetic agent characteristics. In keeping with this philosophy a whole Teletubby was reduced to a polygon count of around only 1200. All agent meshes were then coloured and textured. Given the choice it seems that users prefer agents with reduced polygon models with higher quality textures than models with high polygon counts but textures of only a modest standard.

5.3.3 Virtual Sensors

A fast collision detection component is a necessary aspect of any interactive VE (Cohen *et al.*, 1995). The agent approach to Teletubbyland meant that each individual agent handled its own proximity and collision detection, unlike most VEs that have one overall collision detection component calculating the position of every object within the VE. Traditional collision detection techniques are notoriously slow, as the number of collision tests is $[(n - n^2)/2]$, where n is the number of objects that have to be tested; hence the problem is $O(n^2)$.

The virtual Teletubbies have two invisible bounding volumes used as sensors for obstacle detection and collision. Each agent has an axis-aligned bounding box around it; this is the inner box shown in Figure 5.5. Axis-aligned boxes and spheres are popular choices in VEs due to their simplicity for checking volumes for overlap (Klosowski, 1998). There is also an outer box (not shown to scale) extending five metres in the case of the Teletubbies (and proportionately less for the Noo-noo agent) and acts as a first pass proximity sensor. Any objects intersecting with the outer volume would send a weighted signal to the behaviour-based control architecture, signalling how close they are to an obstacle.

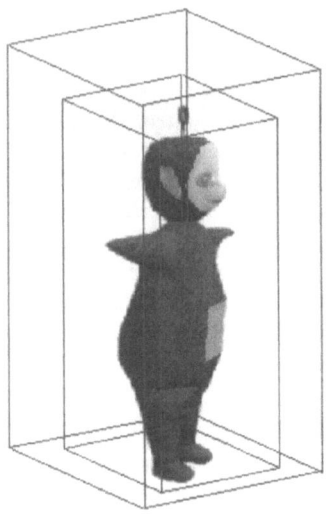

Figure 5.5 Bounding volumes: collision and proximity sensors.

Figure 5.6 Proximity box intersects with hills.

The inner box acts as a collision detection box; any object intersecting it is regarded as touching the Teletubby.

The outer bounding box can be used for agent–agent perception. When these boxes collide with each other, the agents involved "know" that they are in proximity to another agent, and release information to each other about their identity and status. This is intended to add the more explicit coordination needed for group behaviours, such as hugging, that need further development. This is not, of course, what happens for real-world agents, who must derive all information by processing incoming sensor data. However, this "visual system" has proved adequate so far.

There are various problems associated with using bounding boxes for proximity detection, particularly when your agent is covering rough terrain. This is highlighted in Figure 5.6, where Virtual Tinky Winky is shown on the hill (quite common in Teletubbyland); the proximity box is also shown. In this case the proximity box intersects with the ground behind it; this will send a stimulus to the behavioural control system that the agent is quite close to something behind it. In this case this is not a big problem, as Teletubbies only walk forward. However, the Teletubby is prevented from turning round and walking up the hill. An alternative solution might have been a bounding sphere shell, as this would not intersect the gradient of the ground. It would also have sensed an even volume of space, but with the software being used at the time spheres were not possible.

Another problematic situation is when one of the virtual agents is moving down the middle of a thin valley. Now you have the proximity box intersected from both sides by the terrain. The agent architecture receives information that there is an object close to it from the left and then from the right and this continues. This can cause the agent to move forward very slowly and "dawdle", turning left and right. It might possibly not move forward at all and just keep turning on the spot.

To address this problem, in addition to the proximity bounding box it was decided to add a forward frustum sensor that was placed at waist height on the agent, shown in Figure 5.7. The frustum acts as a "ray-scanner" carrying out five sweeps a second, calculating intersections. The combination of the proximity bounding box and front frustum stops the agent from getting trapped.

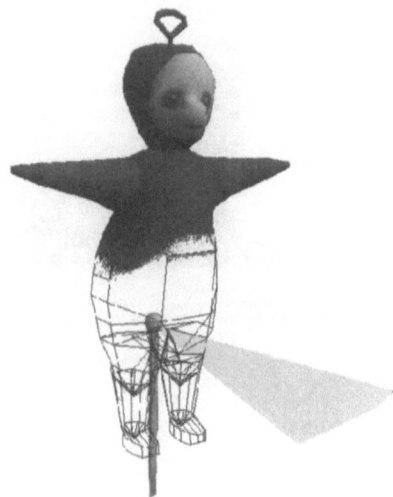

Figure 5.7 The virtual sensors, a front frustum and a ground sensor.

We want the Teletubbies to grasp objects, so the next evolution in the developing the sensors is to add local virtual sensors, possibly taking the semi-automatic approach based on Magnenat-Thalmann *et al.* (1988), with a simple sensor on each joint of the hand.

As well as the frustum sensor an invisible vertical vector sensor is attached to the geometry of the agent. This is continuously detecting the intersection with the ground, and is used for terrain following.

The sensors, which drive the architecture, produce data from both the external environment and the internal state of the agent. Sensor abstraction has been implemented, so that a developer has a common interface in the architecture for all sensors and could expand the library of sensors with more as needed.

5.3.4 Physics in Teletubbyland

A unique problem that you have when creating life-like virtual characters not typically found in robotics is that you have to situate them in an environment that exhibits physical properties, in particular a system of dynamics.

The agent-engine has extra functionality built in for connecting external modules to it; this was designed to make it easy to create routines specific for any other agent environment. One instance of the module subclass, the environment module, was used to create a physics model for the Teletubby world.

A consequence of applying a behavioural architecture is to focus attention on the interaction between agent and environment. Thus it made sense to give the world physical properties apart from the basic one of shape with which the agent can interact. This also meant that other artefacts in the environment obeyed the same laws of physics as the rest of the environment.

The mechanics of the Teletubbyland VE is based on classical Newtonian dynamics. Gravity is based on Newton's second law of motion, which states that the rate of change of momentum of a body is proportional to the resultant force and occurs in the direction of the force, stated as $F = mg$, where F is the force that attracts the object back to Teletubbyland (weight), m is the mass, and g is the acceleration due to gravity (g in Teletubbyland is a unique value not like that found in the natural world). Building gravity into the VE physics engine means that the agent architecture is kept generic and staying on the ground is a property of the world and not something hard-coded into the agent. This also means that when a user enters Teletubbyland as an avatar they are restricted to the same physical constraints of the environment. Without these constraints the user would walk through walls and fly over the land, destroying their experience of being in a Teletubbyland that they think is "real".

5.3.5 The Foundation Architecture

The behaviour-based control system used in the Virtual Teletubbies (BALSA) is an extension of the Behavioural Synthesis Architecture – BSA (Barnes, 1993) that we originally developed so that mobile robots could cooperate to perform tasks such as relocating an object.

The BSA laid the foundation for the success of the project MACTA (Barnes and Aylett, 1994), designed for letting cooperative autonomous mobile robots carry out more complex tasks. Figure 5.8 shows two mobile robots called Fred and Ginger carrying a pipe together navigating round a maze; note the capture-heads on each robot (Ghanea-Hercock and Barnes, 1996): essentially an instrumented X-Y table designed so that the relative motion of one robot would be "communicated" to the other robot and vice versa. One of the main applications that the MACTA project addressed was the use of autonomous mobile robots for automating some parts of decommissioning industrial plants. The previous approach to this problem involved a human operator collectively controlling robots with multiple degrees of freedom from a remote safe location. This was problematic, as coordinating the interactions between the robots became far too complex a task in a safety critical environment.

Figure 5.8 Fred and Ginger performing a cooperative task together.

5.3.6 Real and Virtual Architectures for Embodied Agents

The selection of a suitable architecture for intelligent virtual agents is an inexact science at this stage. Attempts at developing a methodology for the creation of believable agents (Reilly, 1997; Loyall, 1997) have been more oriented to requirements analysis, of both visual appearance and outward behaviour. One can, however, distinguish a spectrum of agents according to whether their main focus is high-level – usually natural-language based – interaction with the user, with cognitive abilities of planning and reasoning, or, at the other end of the spectrum, you have agents who interact with their environment in a physically competent and believable manner with little or no use of cognitive abilities or verbal communication. At the more cognitive end of the spectrum, virtual agent architectures have been based around an affective framework: for example, the Affective Reasoner (Elliot, 1997) and Lyotard (Bates, 1992). Existing architectures from symbolic AI such as SOAR have also been used. Agents who require convincing physical motion have often drawn on robot architectures.

The Teletubbies may be set at the more physical end of the spectrum – they have very limited natural language capabilities and the television series uses stories which depend on the kind of physical interaction with the world appropriate to the toddlers the series is aimed at. An architecture such as that used by STEVE (Johnson *et al.*, 1998), based on SOAR, was therefore not what was required. More appropriate was a behaviour-based architecture, in which virtual sensor inputs are coupled tightly to virtual actuator outputs. Thus a "horizontal" architecture closer to that used with our robots in MACTA met the requirements.

A bottom-up behavioural approach also allows for greater complexity and overall emergent behaviours to arise where an agents behaviour should be more "interesting" and "believable" (Reilly, 1997) for the user, as opposed to top-down rigidly scripted interaction.

Readers looking for a survey paper on Intelligent Virtual Agents are directed to (Luck and Aylett, 2000); this also includes a comparative study of the Virtual Teletubbies with other models.

5.4 Behavioural Architecture for Lifelike Synthetic Agents – BALSA

BALSA is built up of two behaviour sub-modules (Figure 5.9). One of the sub-modules deals with the low-level second-to-second processing, synthesising utility-weighted concurrently active behaviour patterns and carrying out motor control. The other deals with higher level sequencing of behaviours and is predominantly driven by internal motivations. The former is primarily the same as the BSA and the latter is a new module and unique to BALSA. Here each agent runs on a separate thread and can have its own set of behaviours.

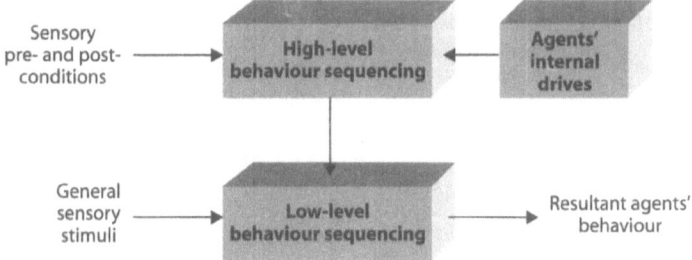

Figure 5.9 BALSA is built of two separate behaviour sub-modules.

5.4.1 Smooth Adaptation Module

Nature has inspired many good pieces of computing (Ballin and Ghanea-Hercock, 2001), and BALSA has drawn inspiration from nature in its control architecture. The low-level processing sub-module deals with the second-to-second activities that the agent will receive, such as obstacle avoidance. The minute-to-minute sub-module deals with sequencing of together of behaviours. The high-level sub-module will feed down relevant behaviours – such as get food – which will be synthesised together with second-to-second behaviour. In effect, the high-level sub-module helps change the mode of the agent smoothly and gets it to do the right thing; we call this *smooth adaptation*.

Nature is littered with example of smooth adaptation. Normally the female and male herring gull take turns incubating the eggs in the nest, whilst the other gets food (McFarland and Bosser, 1993). A sitting bird will not leave the nest until relieved by the partner. However, if the partner is delayed for some unknown reason what does the nesting bird do? Should it wait for its partner to return soon or leave the nest at the mercy of predators and get food? Eventually the gull will change from nesting mode into get food mode. This is in effect what the high-level sub-module does: it gradually changes mode for a behavioural control system, depending on the current circumstances.

5.4.2 Low-level Behaviour Processing

In the BSA, four behaviour levels (often known as strategy levels) were identified for purely conceptual convenience (Barnes, 1996). BALSA has taken the concept of behavioural levels further and actually implemented it into the virtual agent architecture.

A **Self** level contains those behaviour patterns concerned with the maximisation and replenishment of internal resources, e.g. making sure the Teletubby does not go hungry or does not walk up hills when energy is low.

An **Environment** level contains behaviour patterns associated with activities involving objects within the agent's environment, e.g. collision avoidance, collision detection or playing with a toy.

A **Species** level contains those behaviour patterns associated with cooperant activities, e.g. Teletubby group hugging.

A **Universe** level contains behaviour patterns specific to a particular task, e.g. navigating to the initial location of an object to be relocated (such as the location of a slice of bread), then subsequent navigation to the desired goal location (typically the Teletubby toaster).

Sensory stimuli from the agent's sensor systems provide the appropriate internal and external state information needed for the various behaviour pattern levels. From each relevant level appropriate responses are generated that relate to the desired actuation. Any strategy level can contain a number of *behaviour patterns* (*bp*), where:

$$bp = \begin{bmatrix} r \\ u \end{bmatrix} \tag{5.1}$$

and

$$r = f_r(s) \tag{5.2}$$

$$u = f_u(s) \tag{5.3}$$

r is the desired response and is a function, f_r, of a given sensory stimulus, s. Associated with every response is a measure of its utility or importance, u. This quantity is a function, f_u, of the same sensory stimulus. Game theoretic studies (Von Neumann and Morgenstern, 1953) showed that single- and n-player games could be used to model agent versus nature and agent1 versus agent2 in competitive and cooperant scenarios. Strategy selection in these games is dependent upon the information a player may have regarding their opponent's or partner's move and the relative utility (or payoff) of any counter or cooperative move. As this information is analogous to the sensory stimuli available to an agent (virtual or real) and utility is used to great effect in the selection of an appropriate strategy from a set of possible strategies, it was realised that such a concept could be incorporated within our control architecture. Hence a *bp* defines not only what an agent's response (typically motion) should be for a given sensor input, but also provides a measure of how the relative importance of this response varies with respect to the same sensor input. The values of r and u constitute a vector known as a utilitor. Figure 5.10 shows an example of a simple *bp* that might exist at a given level.

Consider the situation where the sensory stimulus relates to an agent's forward-facing distance-to-obstacle measuring sensor and the associated motion response relates to the forward translate velocity for that agent. From Figure 5.10 it can be seen that as the Virtual Teletubby gets nearer to the object then its forward translate velocity will be reduced to zero. At the same time, the associated utility for this motion response increases. Thus as the agent gets nearer to an object in its path, it becomes more important for the agent to slow down. At any point in time, t, multiple conflicting motion responses are typically generated.

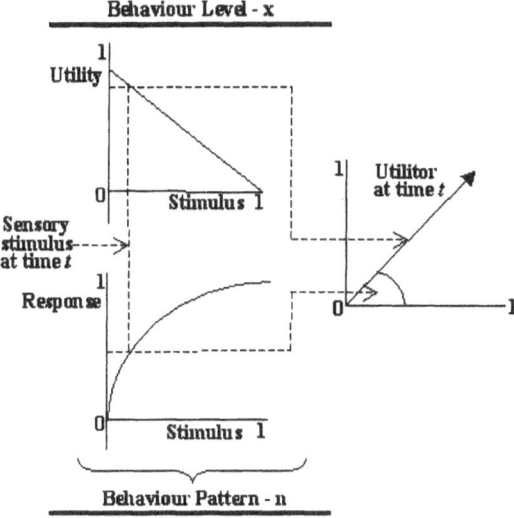

Figure 5.10 Makeup of a behaviour pattern.

For example, a Virtual Teletubby may be navigating towards an object it is curious about when an obstacle unexpectedly appears in its path, at the same time it senses it must go to the Dome and eat, and also that another agent is in the vicinity trying to start a dialogue. In such a situation, what should it do? In the BSA, conflicting motion responses are resolved by a behaviour synthesis mechanism to produce a resultant response. Competing utilitors are resolved by a process of linear superposition which generates a resultant utilitor, UX_t where:

$$UX_t = \sum_{n=1}^{m} u(t,n)e^{j \cdot r(t,n)}$$ (5.4)

and m equals the total number of related utilitors generated from the different behaviour levels; e.g. all those concerned with translation motion or those concerned with rotation motion. Given a resultant utilitor, a resultant utility, uX_t, and a resultant motion response, rX_t, are simply obtained from

$$uX_t = \frac{|UX_t|}{m}$$ (5.5)

and

$$rX_t = \arg(UX_t)$$ (5.6)

X identifies the relevant degree of freedom, e.g. translate or rotate, and the resultant motion response, rX_t, is then executed by the agent. From Equation (5.4), it can be seen that generating a resultant utilitor from different behaviour patterns within the architecture constitutes a process of additive synthesis (Figure 5.11).

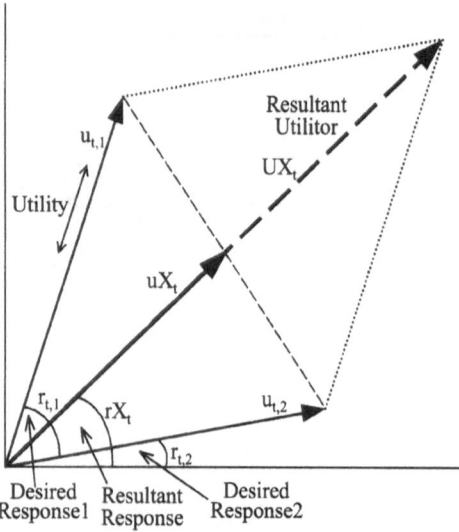

Figure 5.11 Generating a resultant utility and response from *two* constituent utilitors.

If all the *bp*s in an agent's repertoire were active at the same time then the overall emergent behaviour of the agent might be of little value. For example, patterns designed to produce obstacle avoidance – typically those which slow the agent down and make it turn away from an obstacle – are not useful if you want a Teletubby to sit down on a chair or hug another one of its species. The *bp* designer must always bear in mind that the low-level architecture is sensor-driven, and not task- or even sub-task-dependent. What is needed in this case is an automatic mechanism for deactivating the "obstacle avoidance" *bp*s when the "sit" *bp*s or "hugging" *bp*s are active. Associated therefore with every *bp* within an agent is an "active flag", which enables or disables it. Thus obstacle avoidance *bp*s, for example, can be turned off and on when required. A *bp* is "deactivated" in the BSA by forcing the respective utility to zero. The action effectively produces a *bp* of zero importance and hence one which does not contribute to the overall emergent behaviour of the agent.

This mechanism is applied by grouping together *bp*s in goal-achieving sets known as behaviour packets. A behaviour packet is a small data structure which includes a sensory pre-condition for activating the *bp*s it references, and a sensory post-condition which controls deactivation of the named *bp*s. Behaviour packets show some similarity with AI production rules (Davis and King, 1977), though they work at the sub-symbolic level and are driven by incoming sensor data rather than by an inferencing system. They support behavioural sequencing for agents performing at a task (universe) behaviour level.

Behaviour packets can be strung together into a structure known as a *"behaviour script"*, where each packet in turn automatically activates and deactivates the *bp*s it references – those appropriate to the sub-task being carried out (Barnes *et al.*, 1997). A behaviour script thus consists of a set of behaviour

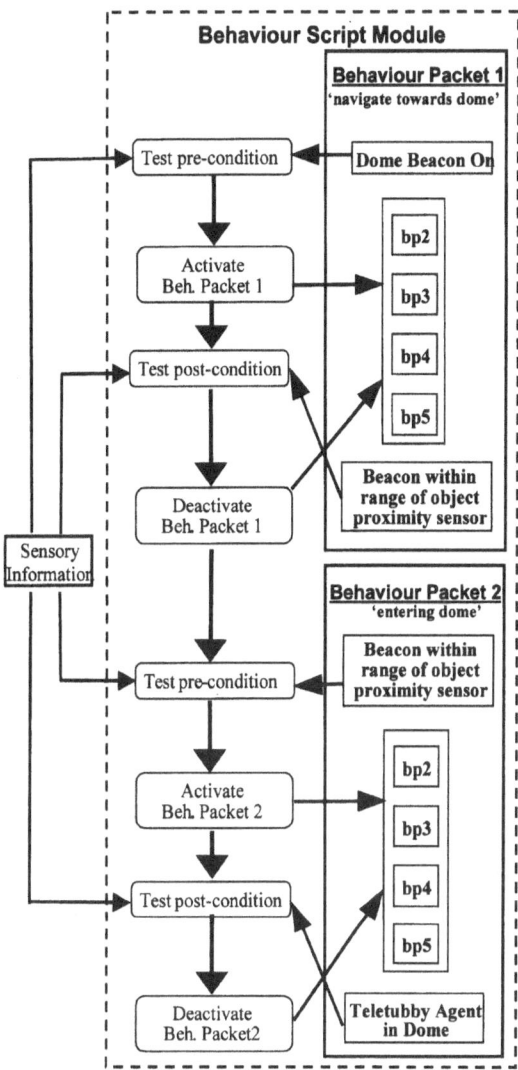

Figure 5.12 Fragment of behaviour script showing a Teletubby entering the Dome. *bp2–bp5* represent appropriate behaviour patterns.

packets, where each forms a triplet: {sensor preconditions, active behaviour patterns, sensor post-condition(s)}. Each script chains together a collection of packets that the agent needs to perform a task.

5.4.3 Motor Control

A motor layer was created which sits in the low-level sub-module of the BALSA architecture. It is used to store movements, like walking, running or reaching, for each kind of Teletubby (Delgado and Aylett, 2000). While behaviour patterns control the direction and speed of motion, limb movement depends on motion capture rather than inverse kinematics for the reasons described above.

In general, this is done by selecting key joints in the desired animal (Gray, 1968) – in our case using a human walking gait and a running gait (Muybridge, 1887). We then digitised each joint to motion capture the data to create an animation of the required movement through the number of desired frames. This technique is called *rotoscoping* and has been used in the computer games industry since the seminal titles *Karateka* and *The Prince of Persia*, both by Jordan Mechner (Red Orb). Later on we used that data to compute the angle of each joint in relation to its parent.

The Teletubby body is stored in a file containing data on the limbs and their relation to the parent joint. The hips are always taken as the root joint. So in a two-legged animal, typically the thighs depend on the movement of the hips, the shins depend on the position of the knees, and the knees depend on the position of the hips. Finally the foot depends on the position of the ankle and the ankle is related to the position of the knee.

A further file is used to store the animation which is carried out using forward kinematics; that is, we first define the position of the hip and then we calculate the orientation of the thigh (and knee joint). Then, with the orientation of the knee we can compute the position of the shin (and ankle). With this approach it is straightforward to incorporate new movements or even to change the bodily form of the agent, without having to start from scratch.

As the Teletubby senses the environment it sends a response as described above, which is converted by the Motor Layer into animation. Each Teletubby has a library of movements, and each Teletubby has a state recording which movement it is currently performing – typically walking.

5.4.4 High-level Behaviour Sequencing

It is possible to generate behaviour scripts on the fly using a reflective agent incorporating a symbolic AI planner, and then send the individual scripts to behaviour-based agents. This hybrid approach was taken with the cooperative robots in MACTA (Aylett, 1996). However, no planning agent is as yet used in BALSA, as this was not deemed necessary: the Virtual Teletubbies are not yet task-orientated enough and it would slow the real-time system down, stifling the user's experience. Instead, a novel sequencing engine was used, linking behaviour to internal motivations or drives.

The high-level behaviour module (Figure 5.13) was developed specifically for BALSA and is not found in the BSA. This also features the internal drives that the virtual agents have, and is comparable to the homoeostatic variables found in work of Bruce Blumberg (1996).

There are several internal drives and imperatives, including hunger, excitability, happiness, curiosity and sleep (Aylett, 1999a; Ballin and Aylett, 1999), though only hunger and sleep are fully implemented as yet. Hunger is the main

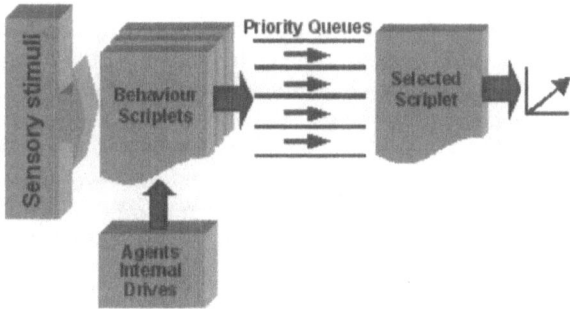

Figure 5.13 Behaviour sequencing.

drive, and increases gradually over time. Once a Teletubby is hungry it will go to the Dome to get Teletubby custard or toast. The user can hold back the Teletubby, but eventually the urge will be too great and it will break free. Currently the Teletubbies do not die, as this was thought to be hard on the children; however, if deprived of food they become lethargic and uninteresting. The sleep imperative is a function of the hunger and energy value – the Teletubbies tend to want to sleep after they have eaten or after long periods of play. Sleep never lasts long, as this would stifle the experience, We are exploring the implementation of different characteristics for each Teletubby – Tinky Winky is characteristically dopey and tends to sleep more; as Dipsy eats more he sleeps often, although for shorter periods, and his personality when awake is quite bold.

In MACTA the robots operated in a single-task environment which produced clear goals, such as "go to a docking beacon". However, the open-ended environment of the Virtual Teletubbies effectively requires a continuous script. Therefore an extra module was created that dynamically generated scripts by interaction with the VE and changes in the agent's internal state.

The framework developed contains four queues, one for each of the conceptual categories *self*, *species*, *environment* and *universe*, as discussed earlier. The entries in this queue consist of groups containing one or more behaviour packets, effectively sub-scripts known as *behaviour scriptlets*, each with an attached priority. The priority is generated automatically and is typically related to a predetermined threshold level of a drive, so the more hungry a Teletubby the greater the priority. The scriptlet with the highest priority is then selected for packet execution. Although the behaviour-sequencing engine is always processing stimuli, it might not be executing a scriptlet, and therefore it has a default script that is at the *environment* level. The default script executes a single packet containing *bp*s that effectively lets the low-level module handle wandering in the environment while avoiding obstacles. The default script is changed when another sensory precondition from another set of packets is met. This is typically at another strategy level, so if a Teletubby sensed the presence of another Teletubby this could trigger the behaviour at the *species* level, while the windmill rotating could trigger behaviour at the *universe* level by drawing all Teletubbies to its vicinity.

Take the following simplified scenario considered during design of the framework: a Virtual Teletubby gets hungry due to the hunger imperative increasing beyond a certain threshold. This behaviour gets prioritised above all others and the Teletubby executes a script which satisfies the goal of going to the Dome to eat. On the way to the Dome the Teletubby still has to avoid obstacles. This is handled by the second-to-second processing of the low-level module. The high-level goal always remains active until satisfied. If another scriptlet gets prioritised above the current one in the stack it will have to wait until the prioritised goal gets executed. Scriptlets always execute sequentially.

What came out of the development of the high-level behaviour module was a sensor-driven sequencing engine. Each scriptlet is effectively a small bunch of packets specifying what the agent should do (as opposed to how). After these groups of scriptlets have executed you have sequenced a small chain of events. So it would be possible to have a Teletubby perform: dock at Dome – get bread – put in toaster – wait till toast ready – eat toast. All of this follows a sensor – action selection – monitor routine. The agent's internal drive and the state of the environment mean that the sequence of events for an agent is frequently different. The sequence of events is less predictable in the case of n agents, where continual interactions with the environment (including other agents) give rise to interesting emergent behaviours.

5.4.5 Embodying Your Television Character

In the early version of the system the only way to observe the Virtual Teletubbies was as a third party. This could be either fully immersed VR (using a headset and gloves) or desktop VR (using a mouse and screen). The desktop VR in some respects has the advantage that you observe at a distance more what was going in the world, and zoom in to where interesting things are happening. The fully immersed VR has the drawback that you miss the bigger picture of the interactions going on, such as not seeing the windmill rotate, and hence you had no idea why the Teletubbies were running toward the hill. Watching from the desktop also had the benefit that you could interfere with the world using a mouse; such as turning and moving Teletubbies around to see if they would avoid one another. It was also possible to raise them from the ground and drop them, which had the surprising emergent effect of making them bounce.

Later versions also allow you to take over the role of one of the Teletubbies and use it as an avatar. The user is now under the constraints of the physics of the virtual environment and cannot fly around but only walk. Now you are recognised by the Teletubbies as one of them, and they are sometimes attracted to you. They will not avoid you, but will avoid colliding with you.

5.5 Future Work

Much implementation of the architecture yet remains to be carried out. But already it is clear that this project has great scope for further developments, and the authors' wish list of future developments is long. For example, motor control and kinematics might be improved, perhaps by using a neural net to cut down the processing overhead (so that the motor control can "predict" the joint movements). Also, improved sensors and a deliberative planner might be added in a hybrid architecture fashion, such as in the MACTA, so that the virtual agents can perform more cognitive tasks, without losing the benefits of behavioural control.

However, a key issue that emerged was that if a user really wants to inhabit a virtual television show a flowing narrative has to be added, in which the user plays a part. Our next key challenge for the project is to create a mechanism so that an *emergent narrative* arises. The idea of interactive drama is not new (Bates, 1992; Hayes-Roth and Brownston, 1995; Perlin and Goldberg, 1996). However, traditional top-down narrative approaches prescribe story structure in advance and therefore limit the degree of freedom that the user experiences in the VE. This removes the feeling of presence, which is one of the key benefits of using a VE in the first place. This demonstrates the narrative paradox of VEs – that the requirement for narrative structure compromises the characteristics that make a VE different from other media. This has given rise to our concept of an emergent narrative arising through *social presence,* something we are keen to develop and add. An overview of this and other narrative approaches for VE's can be found in Aylett (1999b).

On analysing some of the Teletubby stories on television, a very simple structure emerges. Narrative normally hinges on an environmental disturbance and the reaction of characters to it. For example, in one story, one of the Teletubbies was overcome by an uncharacteristic urge to run around shouting "Run, run, run". As they passed by the other Teletubbies, each in turn was overcome by the same urge, until all four were running along in a line. In another story, a large pair of rubber boots suddenly appeared in Teletubbyland. The Teletubby that came across the boots put them on and stamped around in them. A second Teletubby was attracted and was given the boots to put on, but promptly fell over as soon as it tried to walk, because it was smaller and the boots didn't fit. Both of these stories could plausibly be produced without pre-scripting as long as a sufficiently rich repertoire of behaviours is implemented for each Teletubby and an event generator is included to provide environmental disturbances.

Behaviours supporting physical interaction of a more interesting kind, particularly at a species level where agent interaction occurs, need to be added. More "physical" activities, such as picking objects up or hugging, currently have lots of problems. As one would expect in a sensor-driven architecture, this is predominately due to impoverished sensing, especially poor tactile sensors with no haptic feedback. Successful hugging requires that Teletubbies avoid crushing each other while still making proper contact. Neither is it desirable to have arms going through necks or bodies in a physically implausible manner, though without much improved collision detection this is all too likely.

A further source of interesting interaction, apart from better physical behaviours, would be the implementation of a richer emotional model in the

Teletubbies. In particular, as the "Run, run, run" story demonstrated, agents who have little verbal interaction need emotional interaction to stimulate behaviour. The ability to propagate more interesting emotional states between Teletubbies is one more area for further investigation.

5.6 Conclusions

The development of BALSA proved a relatively straightforward exercise. Greater problems were experienced with the virtual sensors and some degree of experimentation was required with their level of sensitivity. It remains a challenge to develop a debugging tool that allows developers to see whether problems are with the agent's sensors, motivations or behaviours.

This chapter has concentrated on the behavioural architecture of the Virtual Teletubbies, highlighting how the long-term goals that are generated are also intertwined with the low-level reactive behaviours in real time. This mechanism also provides the synthetic characters with a smooth transition between action-selection of key behaviours. Innovative work on this project has also taken place with the development of the physics engine and virtual sensors (Ballin and Aylett, 1999).

The project is not yet at a stage where you can become a television character and be totally integrated into the artificial world (although it has now reached a stage where you can be entertained). VE may be a new medium, but mapping television and films to it is not straightforward. But we feel we have made headway, and this approach, coupled with an emergent narrative, will take us down a road where maybe everybody can be a superhero for a while.

Acknowledgements

Thanks to the original Virtual Teletubby team who transformed a 2D idea into the illusion of life. Also thanks to Ragdoll Productions for allowing us conduct this research. The Teletubby characters are copyright © 1996 Ragdoll Productions (UK) Limited. Batman is copyright © of DC Comics, and Luke Skywalker is the property of Lucasfilm Ltd. However, the authors take full responsibility for the "Virtual Teletubbies".

References

Aylett, R. S. (1996) Communicating goals to behavioural agents – a hybrid approach. In *Proceedings of the Third World Congress on Expert Systems*, Seoul, February.

Aylett, R. S. (1999a) Behavioural virtual agents. In *Artificial Intelligence Today* (eds. M. J. Woolridge and M. Veloso). Springer-Verlag, pp. 1–11.

Aylett, R. S. (1999b) Narrative in virtual environments – towards emergent narrative. *AAAI Fall Symposium*.

Ballin, D. and Aylett, R. S. (1999) Exploring a behaviourally-driven architecture for 3D autonomous virtual agents. *Technical Report*, Centre for Virtual Environments, University of Salford.

Ballin, D. and Aylett, R. S. (2000) Time for Virtual Teletubbies: the development of interactive autonomous children's television characters. In *Workshop on Interactive Robotics and Entertainment (WIRE-2000)*, Robotics Institute, Carnegie Mellon University, 30 April–1 May, pp. 109–116.

Ballin, D. and Ghanea-Hercock, R. (2001) Back to nature. *The British Computer Society Computer Bulletin* Series V, **3**(2).

Barnes, D. P. (1993) Multiple co-operant mobile robots for advanced manufacturing applications. *EPSRC Grant GR/F71454.*

Barnes, D. P. and Aylett, R. S. (1994) ITE: multiple automata for complex task achievement. *EPSRC Grant GR/J49785.*

Barnes, D. P. A behaviour synthesis architecture for cooperant mobile robots. In *Advanced Robotics and Intelligent Machines* (eds. J. O. Gray and D. G. Caldwell). IEE Control Engineering Series 51, pp. 295–314.

Barnes, D. P., Ghanea-Hercock, R., Aylett, R. S. and Coddington, A. M. (1997) Many hands make light work? An investigation into behaviourally controlled cooperant autonomous mobile robots. In *Proceedings of the first International Conference on Autonomous Agents* (ed. L. Johnson). Marina del-Rey, CA, ACM Press, pp. 413–420.

Bates, J. (1992) Virtual reality, art and entertainment. *Presence: The Journal of Teleoperators and Virtual Environments* 1(1), 133–138.

Benford, S., Greenhalgh, C., Brown, C., Walker, G., Regan, T., Rea, P., Morphett, J. and Wyver, J. (1998) Experiments in inhabited TV. In *CHI'98 Late Breaking Results* (Conference Summary). ACM Press, pp. 289–290.

Blumberg, B. (1996) Old tricks, new dogs: ethology and interactive creatures. *Ph.D. Dissertation*, Massachusetts Institute of Technology.

Boulic, R., Capin, T. K., Huang, Z., Kalra, P., Lintermann, B., Magnenat-Thalmann, N., Moccozet, L., Molet, T., Panzic, I., Saar, K., Schmitt, A., Shen, J. and Thalmann, D. (1995) The HUMANOID environment for interactive animation of multiple deformable human characters. In *Proceedings of Eurographics '95*, Maastricht, August, pp. 337–348.

Cohen, J., Lin, M., Manocha, D. and Ponamgi, K. (1995) I-collide: an interactive and exact collision detection system for large-scale environments. In *ACM International 3D Graphics Conference '95*, pp. 189–196.

Davis, R. and King, J. (1977) An overview of production systems. In *Machine Intelligence* (eds. E. W. Elcock and D. Michie). New York, Wiley & Sons, pp. 300–332.

Delgado, C. and Aylett, R. S. (2000) From robots to creatures: extending a robot behaviour based architecture to create virtual creatures. In *Proceedings, ISRA 2000*, Monterrey, Mexico.

Elliot, C. (1997) I picked up Catapia and other stories: a multimodal approach to expressivity for "emotional intelligent" agents. In *Proceedings of the First International Conference on Autonomous Agents* (ed. L. Johnson). Marina del-Rey, CA, ACM Press.

Ghanea-Hercock, R. and Barnes, D. P. (1996) Coupled behaviours in the reactive control of cooperating mobile robots. *International Journal of Advanced Robotics, Japan*, **10**(2), 161–177.

Gray, J. (1968) *Animal Locomotion*. London, Weidenfeld & Nicholson.

Hayes-Roth, B. and Brownston, L. (1995) Multiagent collaboration in directed improvisation. In *Proceedings of the First International Conference on Multi-Agent Systems (ICMAS-95)*, San Francisco, CA, pp. 148–154.

Johnson, W. L., Rickel, J., Stiles, R. and Munro, A. (1998) Integrating pedagogical agents into virtual environments. *Presence: Teleoperators and Virtual Environments*, 7(6), 523–546.

Klosowski, J. T. (1998) Efficient collision detection for interactive 3D graphics and virtual environments. *Ph.D. Thesis*, Department of Applied Mathematics and Statistics, State University of New York at Stony Brook.

Loyall, B. A. (1997) Believable Agents. *Ph.D. Dissertation*, Carnegie Mellon University.

Luck, M. and Aylett, R. S. (2000) Applying artificial intelligence to virtual reality: intelligent virtual environments. *Applied Artificial Intelligence*, **14**, 3–28.

Magnenat-Thalmann, M., Laperrière, R. and Thalmann, D. (1988) Joint-dependent local deformations for hand animation and object grasping. *Proceedings of Graphics Interface '88*, Edmonton.

McFarland, D. and Bosser, T. (1993) *Intelligent Behaviour in Animals, and Robots*. Cambridge, MA, MIT Press/Bradford Books.

Muybridge, E. (1887) *Animal Locomotion*. University of Pennsylvania.

Pausch, R., Snoddy, J., Taylor, R., Watson, S. and Haseltine, E. (1996) Disney's *Aladdin*: first steps toward storytelling in virtual reality. In *Proceedings of the Twenty Third International Conference on Computer Graphics and Interactive Technique (SIGGRAPH'96)*, ACM Press, pp. 193–203.

Perlin, K. and Goldberg, A. (1996) Improv: a system for scripting interactive actors in virtual worlds. In *Proceedings of the Twenty Third International Conference on Computer Graphics and Interactive Technique (SIGGRAPH'96)*, ACM Press, pp. 205–216.

Reilly, W. S. N. (1997) A methodology for building believable social agents. *Proceedings of the First International Conference on Autonomous Agents*. Marina del-Rey, CA, ACM Press, pp. 114–121.

Thalmann, N. M. and Thalmann, D. (1998) The virtual human story. *IEEE Annals of the History of Computing*, 20(2), 50–51.

Ventrella, J. (1998) Designing emergence in animated artificial life worlds. In *Virtual Worlds 98, LNAI, 1434* (ed. J.-C. Heudin). Springer-Verlag, pp. 143–155.

Von Neumann, J. and Morgenstern, O. (1953) *Theory of Games and Economic Behaviour*, 3rd edn. Princeton, NJ, Princeton University Press.

Walker, G. (1997) The mirror – reflections on inhabited TV. *British Telecommunications Engineering Journal*, 16(1), 29–38.

Towards an XML and Agent-based Framework for the Distributed Management and Analysis of Multi-spectral Data

Omer F. Rana, Yanyan Yang, Christos Georgousopoulos, David W. Walker, Giovanni Aloisio, and Roy Williams

Abstract

Implementation and use of a prototype distributed active data archive system are outlined. This system is based on the Synthetic Aperture Radar Atlas (SARA) and utilises cooperative software agents for data access and analysis, and uses XML to model metadata and support agent communication. The analysis is undertaken on a digital library containing images of the Earth acquired by the Space Shuttle, and subsequently enhanced by data derived from Earth stations. An application scientist can view an existing image, combine an image with textual data for a particular region from a regional geographic information server, or combine images for a given region over a period of time to look for changes in a region.

6.1 Introduction

The general problem of managing large digital data archives is particularly challenging when the system must cope with active data which is processed on demand (Coddington *et al.*, 1998). Active data represents data that is dynamically generated by a scientific experiment, or may be obtained from a sensor or monitoring instrument. SARA (Synthetic Aperture Radar Atlas) is an active digital library of multi-spectral remote sensing images of the Earth, and provides Web-based on-line access to such images. SARA has been online for over a year at the University of Lecce in Italy, at Caltech in California, and at SDSC in San Diego, California, being primarily maintained by Caltech and the University of Lecce. Scientists making use of the SARA archive often require integrated access to information, combining retrieval, computation and visualisation of multiple images. Such an analysis can involve overlapping images to detect change within a given area or colour coding images based on radar frequencies and the ground characteristics of the region being investigated.

A multi-agent system, which comprises both intelligent and mobile agents, has been developed to manage and analyse distributed multi-agency remote sensing data. Compared with other agent applications, our emphasis is on the use of the mobile agents to transfer analysis algorithms to image processing archives. We localise the most complex functionality in non-mobile Local Interface Agents (LIA), which remain at one location, communicate with the mobile User Interface Agents (UIA), and provide resources and facilities to lightweight mobile agents that require less processor time to be serialised and are quicker to transmit. Each agent is responsible for offering a particular type of service, and the integration of services is based on a user specification. SARA mobile agents are persistent, and can wait for resources to become available. Agents allow the delivery and retrieval of data to complete without user monitoring or recovery actions. There are two types of User Interface Agent: User Request Agents (URA) and User Assistant Agents (UAA). URA supports the user in creating a query or operation to perform on the SARA data. UAAs manage the information of the user and provide control functions to the user, such as updating their file space on a remote server, and parameter settings for their visualisation tool. There are many types of Local Interface Agent: a Local Assistant Agent (LAA) supports interaction with any visiting User Request Agents (URAs) by informing them about the available data and computing resources, and cooperating on the completion of the task carried by the URA. A Local Management Agent (LMA) coordinates access to other LAAs and supports negotiation among agents. It is responsible for optimising the itineraries of mobile URAs to minimise the bottlenecks inherent in parallel processing and ensuring that the URA is transferred successfully. A Local Inte-Gration Agent (LIGA) provides a gateway to a local workstation cluster.

The eXtensible Markup Language (XML) is becoming the standard for data interchange on the Internet, and enables a new generation of Web services that are not meant for humans to use directly, but rather to be used by other software. Every XML document refers to a Data Type Definition (DTD), which is a grammar that defines the document syntax. In our system we use XML to encode system structure as metadata and user requests. When a user launches a query, an XML document is created. Every specific XML specification is based on a separate DTD that defines the names of tags, their structure and content model. While the XML specification contains the structured information, the DTD defines the semantics of that structure, effectively defining the meaning of the XML document. An agent can generate the processing programs representing the XML elements of interest according to the DTD with a parser, and then travel across the Internet to retrieve related information (for example, data corresponding to the tag TRACK in XML).

We also use XML as an application-specific transport protocol to enable agents within the system to communicate with each other. Autonomous agents cooperate by sending messages and using concepts from the SARA ontology –

where the ontology describes terms and concepts (such as a Track or a Latitude/Longitude coordinate) and their interrelationships. We are defining a message which embeds such an ontology to provide a mechanism for agents to exchange requests and message interpreters. Agents send and receive information through XML-encoded messages. Based on pre-defined tags, agents may use different style DTDs to fit different mediation. Moreover, a mobile agent can carry an XML front end to a remote data source for data exchange, where both queries and answers are XML encoded. In our prototype, we use the JAXP (Java API for XML Processing; http://java.sun.com/xml/jaxp/index.html) interface to XML developed by Sun, which supports SAX and the general purpose Document Object Model (DOM). We currently have a working prototype. Our future work is to extend the ontology's definition and investigate system scalability and response times when thousands of agents are hosted on a single data source.

6.2 Types of Operation Supported by SARA

The data maintained within the SARA system was acquired by the Space Shuttle during a week-long flight covering an area of roughly 50 million square km. The data was acquired using "active" radar, which measures the strength and round-trip time of the microwave signals that are emitted by a radar antenna and reflected off a distant surface object. Hence each pixel in the generated image corresponds to radar backscatter. Darker areas in the image represent low backscatter and bright areas represent a high backscatter. Bright features mean that a large fraction of the radar energy has been reflected back, where the backscatter depends on the size of the scattering objects in the target area, the moisture content of the target area, the polarisation of the pulses and the observation angles. The microwave transmissions are vertically and horizontally polarised, and received on two separate channels. Hence radar backscatter can be in four polarisation combinations: HH (horizontally transmitted, horizontally received), VV (vertically transmitted, vertically received), VH and HV. This enables the derivation of the complete scattering matrix of a scene on a pixel-by-pixel basis. Subsequent analysis involves allocating colours to these polarisations to identify particular surface features, such as vegetation cover and sub-surface discontinuities. Additional details of how the imaging radar works can be found at http://southport.jpl.nasa.gov/.

When using the SARA system, a user selects a particular part of the Earth, choosing an area where data is available, shown as rectangles in Figure 6.1. Alternatively, the user can create a polygon over the surface of a globe/map, from which latitude and longitude coordinates are derived. A URL is generated whose content is the multi-channel dataset (image) to be analysed by a particular algorithm. The generated URL corresponds to the region of interest as an SAR image. The processed multi-spectral data may be further processed by choosing a mapping from the frequency/polarisation channels to be red, green and blue components of the final image. This mapping may be optimised to highlight aspects such as ground ecology or snow/ice conditions, for instance. A user server communicates with a metadata server, which contains metadata

Figure 6.1 SARA map for the UK – the rectangles show available tracks.

descriptions such as the position of the image on the surface of the Earth, and the user server opens an HTTP connection to a data server. The data server retrieves the requested data from a mass storage system, and delivers the image as a JPEG file. Additional compute servers are provided for performing activities such as image processing, land classification etc. The architecture of the system is illustrated in Figure 6.2 and obtained from or http:// sara.unile.it/sara/. The data server is linked to an HPSS/Unitree file system, from which the required image to be viewed, or analysed, is retrieved and passed to a compute server. The compute server can also connect to specialised visualisation engines, such as a CAVE environment. The file system can connect to the compute server via a parallel interconnect, as illustrated.

6.2.1 Data Tagging and Integration

The XSIL (Blackburn *et al.*, 1999) representation is used to tag data within the SARA digital library. XSIL is based on the eXtensible Markup Language (XML; http:// www.xml.com/), and used to translate coordinates identified by the user into server locations and filenames. Code segment 1, below, demonstrates a simple XSIL description for SARA resources, where coordinates for a quadrilateral drawn by a user are subsequently passed to an HPSS system for the retrieval of the corresponding image. Since the image repository is replicated, a timeout limit is imposed on retrieving an image from a server, after which an alternative (mirror) server is contacted. For instance, the request passed to the data source (identified by the baseurl in code segment 1) is of the form http://www.cacr.caltech.edu/cgi-bin/sara_metadata_server?lat=52.26&lon=-0.23. One of the primary reasons for using XSIL is to overcome the deficiency with stateless CGI (Common Gateway

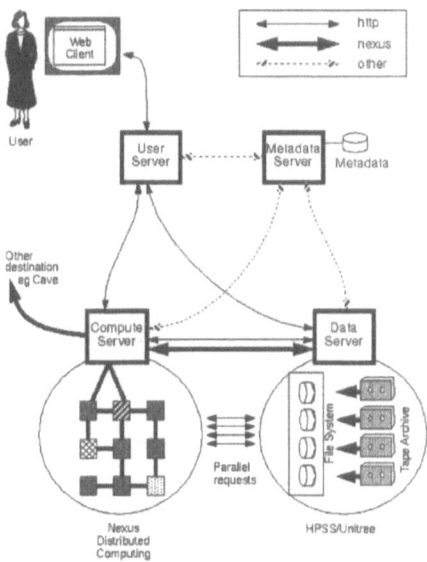

Figure 6.2 Architecture of the SARA system.

Interface) scripts used to access data resources connected to a Web server. Hence with CGI scripts, a user cannot enable one request to be based on the outputs or conditions generated from a previous one. Furthermore, the output from a CGI script is a MIME-typed data object, which restricts the passing of compound multiple data objects in response to a user request.

```
<track id=13242>
  <name>Almaz, Russia</name>
  <date>04-16-1994</date>
  <width>2824</width>
  <height>8000</height>
  <area>
    <quadrilateral>
      <lon>47.425</lon><lat>45.669</lat>
      <lon>47.703</lon><lat>45.418</lat>
      <lon>48.719</lon><lat>45.963</lat>
      <lon>48.444</lon><lat>46.217</lat>
    </quadrilateral>
  </area>
  <server name=CACR_HPSS>
    <baseurl>http://hpss.cacr.caltech.edu/...</baseurl>
    <timeout>660</timeout>
  </server>
  ...
</track>
```

Code segment 1

Each request passed to the data source has a unique identifier, which can subsequently be tracked by the user. The data from a particular request can vary in size from large data objects a few megabytes in size to small data sets generated by more complex requests. Hence a request could be a filename and a simple filter to be applied to that file, or an SQL query to a database whose output is a list of files which satisfy the query.

6.3 SARA and Data Management

One of the problems with using the SARA system is the quantity of data that could be generated during a single request. If the user requesting analysis on the generated data is not at the data source, hundreds of Mbyte of data may need to be downloaded to the client. There are two ways to deal with this:

1. Download this data to a local parallel computer which is presumably closer to the data source than the user, and perform analysis on the parallel machine.
2. Migrate the computation required by the user to the data source, or to a site that is closer the user, and send analysis results back to the user.

As indicated previously, currently approach (1) is adopted, whereby computing resources to use are identified beforehand, and on which the image processing activity is performed. The results are examined using a Web browser – enabling further data to be imported and processed if desired. The user can also create scripts that will enable production runs that can execute for longer periods of time. The image corresponding to a particular region selected by the user can be further combined with information in commercial databases, such as Geographic Information Systems (GIS). When used in this way, landmarks such as towns, roads, rivers and lakes can be superimposed on the image retrieved from the HPSS system. Such an integration of land features into an image generally involves data fusion, rather than data analysis. In this instance, the longitude and latitude information maintained by the metadata server is essential to enable a GIS system to locate information on a region of interest. Hence the SARA system involves:

- *Data Fusion and Integration* supported by the metadata server. In this instance, the metadata server provides summary information, rather than the structure of the data source. The metadata is encoded in XSIL and added to the coordinate information used to locate images of interest, details of servers and time-out periods associated with particular data servers.

  Data fusion can also involve combining SARA images to study the changes in a region of interest over time, or generating multi-perspective views on a given region of interest. In this context, colour coding can be used to isolate certain features of an image, for instance.

- *Data Storage* involves maintaining the images on an HPSS/Unitree file system and replicating the stored images at various places around the world. The time to retrieve the image can vary depending on the request, but is generally of the order of a few minutes. The request can also involve the loading of a data storage unit via a tape robot. The storage system is connect to a Web server via a CGI script.

- Data analysis can involve various different operations, such as edge detection or principal component analysis, or more complex approaches based on neural networks or genetic algorithms.
- The results can be visualised using a Web browser, or via a more sophisticated immersive environment, such as a CAVE.

6.4 Agent-based Prototype

We make use of mobile agents to carry queries to the data source, rather than moving the data generated in response to a query to the data processing site. In our system, we divide operations performed by a user into a number of interacting agents, with each agent delegated a particular role within the system. Each user is presented with an interface agent, which manages user interactions with the local file system and provides support for launching a query to the SAR data. The query is wrapped as a User Request Agent, using the Voyager mobile object library (http://www.objectspace.com/), and delivered to the SARA data server. The agent passes the incoming query to the data server, where the query is executed and any data generated is maintained as a local file. A URL reference of the file is handed back to the User Request Agent, which can then proceed to another host for additional processing, or to the parent host with the URL of the generated data and the status of results that are to be sent to the user. The interface agent at the user side receives the results and displays these to the user. In case of additional processing, the User Request Agent is despatched to additional hosts, and continues to carry the URL of data sources at each intermediate site that it visits. Hence we do not migrate data sets across networks, as the processing demands may vary from a simple change of data format to more compute intensive image processing, such as principal component analysis or pattern classification.

As illustrated in Figure 6.3, the user downloads a Web page containing an Applet from a Java-enabled Web browser (Step 1) and then initiates and despatches a User Request Agent (URA) (Step 2). The URA then takes a user's query request encoded in XSIL to the first stop of its itinerary – SIR-C imaging radar in the USA – and interacts with the Local Security Agent (LSA) (Step 3). After authenticating the sending user through an LSA, the URA interacts with the Local Assistant Agent (LAA) to find out about the available data source and services (Step 4). Once the required resources in the local sites are available, a URA communicates with the Local Request Agent (LRA), and the LRA completes the retrieve task at the local sites and saves the results to a file. Then the URA creates a URL that points to that file (Step 5). The URA then dispatches itself to the next stop on its itinerary and communicates with the agents of the next site – ERS-1 data in Italy (Steps 6, 7, 8). Finally, the URA communicates with the User Presentation Agent (UPA) and returns the URL(s) pointing to the result file(s) to the UPA when the itinerary is completed. The URA then dies

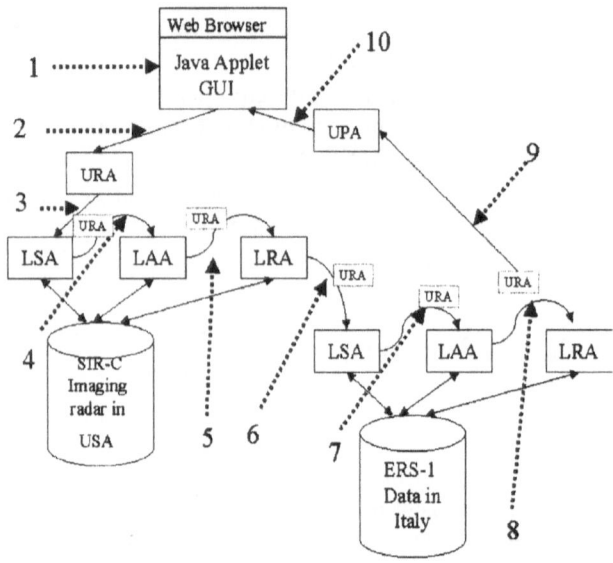

Figure 6.3 A usage scenario of the agent-based SARA system.

(Step 9). The UPA presents the results to the user through the graphical interface according to the rules defined in the UPA (Step 10). In this way, a user can
connect to the network at a low bandwidth while the data archive is connected
to a powerful compute server over a high bandwidth network connection, and
the user can also engage in other tasks or shut down the machine after
despatching the URA. The user interface is illustrated in Figures 6.4 and 6.5 and
enables a new algorithm to be wrapped as an agent and migrated to the
compute server identified previously.

For example, URA1, identified in Figure 6.3, has the following itinerary: visit
server Sl, perform task T1, and store the result data in S1; then visit server S2
and perform task T2, visit server S3 and perform task T3. The merged result
data of T2 and T3 are stored on S3. Next, suppose that after URA1 performs its
work, URA2 is sent out with the itinerary: visit server S1 and perform task T4,
visit server S2 and perform task T2; then visit server S3, and perform task T3.
When URA2 arrives at S1, Local Management Agent 1 (LMA1) on S1 will realise
that the results of T2 and T3 can be found on S3. Rather than send URA2 to S2,
the LMA1 changes URA2's itinerary so that it need not visit S2 and S3, and just
returns the URL reference on S3. Alternatively it could just visit S3 and return
back with the result data.

Hence a URA can retrieve the required image from the SARA data server and
migrate the results to a GIS server to add additional land features. The resulting
composite data can then be migrated to the compute server. A mobile agent
carrying an analysis algorithm is then sent to a compute server, where subsequent image analysis is undertaken. The agent-based approach therefore
divides the operations undertaken into a collection of roles, where each agent
offers a particular service. In this case, the URA is acting as a carrier of code

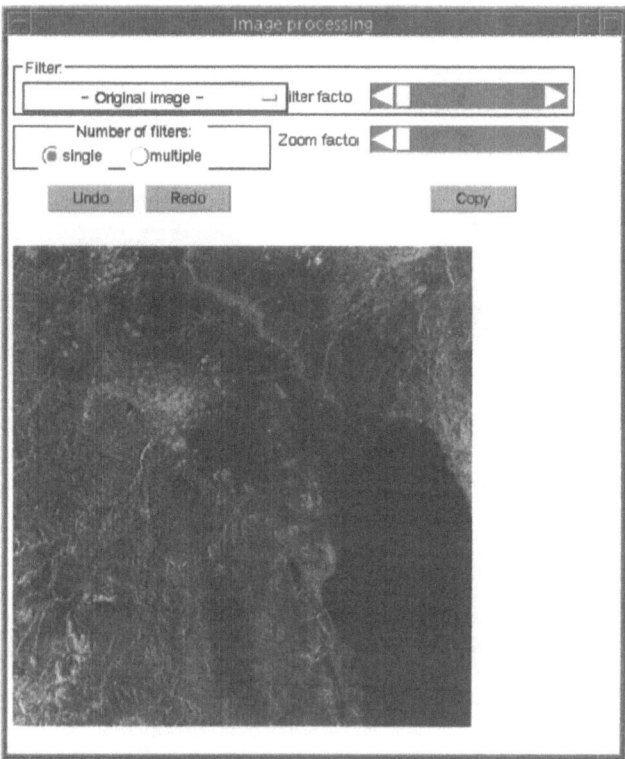

Figure 6.4 The Graphical User Interface (GUI) showing a retrieved SARA image – the colours correspond to polarisation information based on electromagnetic polarisations.

and data, and is interacting with multiple stationary agents, such as the Local Security Agent (LSA) and the Local Request Agent (LRA). In our case, the LSA maintains a list of certificates identifying sites that can send their agents to this particular data or compute server. Similarly, the LRA abstracts the type of data source that contains the data, with the data source ranging from flat files to structured databases. Hence the LRA can either retrieve files from a file system or execute SQL queries via a database management system. This approach therefore provides a "logical" definition of a storage system, where the "physical" storage system can be implemented using any storage technology, such as Unix file systems, HTTP servers, Hierarchical Storage Systems (such as HPSS), and network caches such as the Distributed Parallel Storage System (DPSS). Consequently, we treat the data storage aspect in a similar way to the Data Grid ideas outlined in Chervenak *et al.* (1999).

Agents can therefore provide specialised services within the context of SARA, or can act as information brokers, helping to locate resources or services of interest (Rana and Walker, 2000). The use of an information sharing and agent collaboration scheme reduces the number of costly server requests as well as

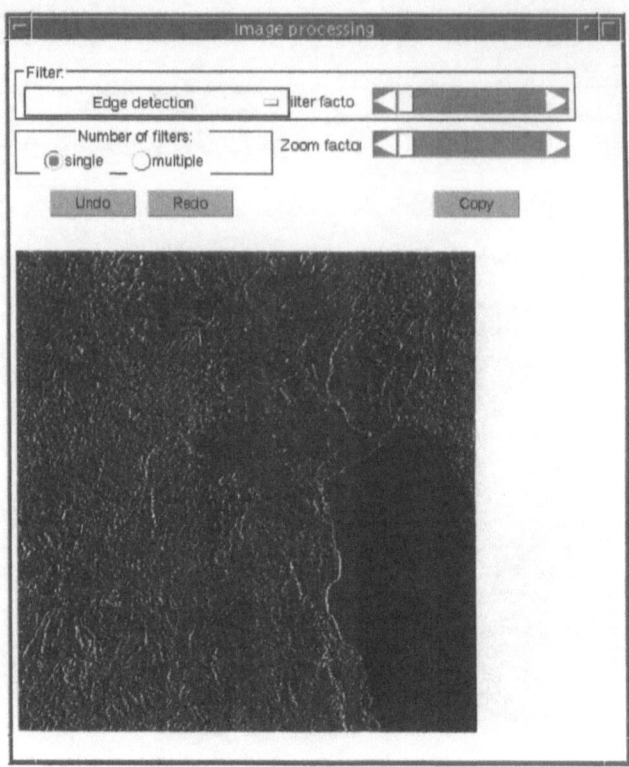

Figure 6.5 The edge detection algorithm applied to a retrieved SARA image.

enabling automatic and dynamic configuration of distributed parallel computing resources. Additional details about the types of agent involved and implementation can be found in Yang *et al.* (2000).

Although the compute servers are still managed and controlled as with the Web-based system, the agent-based approach provides additional flexibility in enabling users to develop their own analysis algorithms. Furthermore, the agent abstract enables a diverse range of data sources, ranging from the original SARA repository to GIS databases, information from ground stations, and data obtained from weather satellites, to be fused. An agent can visit multiple sites on its itinerary, prior to being returned to the user. Similarly, the security agent can support different types of access rights, based on a local access policy of the data or compute server administrator.

6.5 Conclusion and Future Work

The agent-based approach therefore offers the following advantages:

- *Modularity:* the system may be composed of interchangeable modules, each providing some of the required functionality. For example, Local Request Agents (LRAs) can translate a query between different formats, and retrieve information from a local archive, which could be a database system or a file system. If the local archive system changes, we need only change the LRA.

- *Scalability:* the system can deal with a large number of requests coming from many users simultaneously. As the load grows, the system scales gracefully. For example, in our architecture, the Local Management Agent may intelligently assign computational resources. A more thorough investigation of scalability in multi-agent systems can be found in Rana and Stout (2000).

- *Decentralisation:* the system does not contain a global administrator. Hence agents can be submitted from various remote sites and it would be inefficient to route all agents through a central site. Control is therefore distributed.

- *Extensibility:* the system components should allow new elements, such as new services and new archive systems, to be easily added.

The agent-based approach therefore provides an integration of different types of data storage and analysis approaches within a single framework. Agents communicate with each other using XSIL-based messages, and can share both data and code. The current system can be employed in a wide range of application domains, such as the analysis of multi-temporal images corresponding to changes in the ecology of a particular region, for studies of environmental pollution for instance. SARA images can also be compared, based on phase and amplitude differences of the backscatter radiation, to study seismic or volcanic processes, motions of ice sheets or glaciers, or other similar geological events. Support for real-time processing can facilitate frequent over-passing of satellites over a given region in the event of natural disasters such as forest fires or flash floods. The agent-based approach provides a useful system for enabling such applications to be more effectively deployed (for the reasons mentioned above), and involve an integration of data resources, various types of physical storage media and various compute servers and analysis algorithms.

References

Blackburn, K., Lazzarini, A., Prince, T. and Williams, R. (1999) XSIL: Extensible Scientific Interchange Language. In *Proceedings of HPCN'99*, Amsterdam, February, pp. 513–524.
Chervenak, A., Foster, I., Kesselman, C., Salisbury, C. and Tuecke, S. (1999) 'The data grid: towards an architecture for the distributed management and analysis of large scientific datasets. Available from: http://www.globus.org/.
Coddington, P. D., Hawick, K. A., Kerry, K. E., Mathew, J. A., Silis, A. J., Webb, D. L., Whitbread, P. J., Irving, C .G., Grigg, M. W., Jana, R. and Tang, K. (1998) Implementation of a Geospatial Imagery Digital Library using Java and CORBA. Available from: http://www.cs.adelaide.edu.au/.
Rana, O. F. and Stout, K. (2000) What is scalability in multi-agent systems? In *Proceedings of Autonomous Agents 2000*, Barcelona, Spain.
Rana, O. F. and Walker, D. W. (2000) The agent grid: agent-based resource integration in PSEs. In *Proceedings of 16th IMACS World Congress on Scientific Computation, Applied Mathematics and Simulation*, Special Session on Problem Solving Environments, Lausanne, Switzerland, August.
White, J. E. (1996) Mobile agents. *General Magic White Paper*. Available from: http://www.genmagic.com/agents.

Williams, R. and Sears, B. (1998) A High-performance active digital library. *Parallel Computing*, Special Issue on Metacomputing, November.

Williams, R., Bunn, J. and Moore, R. (1998) Interfaces to scientific data archives. *Report of a Workshop sponsored by the National Science Foundation*, May 1998. Available from: http://www.cacr.caltech.edu/isda/.

Yang, Y., Rana, O. F., Georgesouplous, C., Walker, D. W. and Williams, R. (2000) Mobile agents and the SARA digital library. In *IEEE Conference on Advanced Digital Libraries*, Washington DC.

7

Learning by Experience – Autonomous Virtual Character Behavioural Animation

Tao Ruan Wan and Wen Tang

Abstract

In this chapter, we present a novel goal-orientated approach for complex virtual character behavioural simulation. Our approach is based on the concept of an artificial brain that is a simulation of real human brain activities for simulating the virtual character's behaviour in a virtual world populated with other virtual objects and characters. The control unit in the simulation system can collect and store all the information that is obtained through the virtual character's complex experience, such as learning how to walk and jump, and analyses the information to find a better solution for a specific task. Therefore the virtual character's skill in a particular task will be developed or evolved. The core techniques also include a physics-based human model for motion modelling, which is driven by muscle forces. This approach therefore produces a more accurate simulation of the real world than conventional methods. We demonstrate this by presenting an implementation of this approach.

7.1 Introduction

One of the most interesting topics in the computer graphics and animation area has been "bringing to life". The importance of producing lifelike computer animations has been identified from the early 2D cartoons (Thomas, 1994) to the latest real-time 3D applications, such as video games and virtual training (Badler *et al.*, 1999; Cavaza *et al.*, 1998). However, despite the huge efforts that have been made in creating lifelike virtual characters in the virtual world in the last two decades and the great successes that have been achieved in many aspects of the virtual world, the results produced over the years are still far from reaching our ultimate goal.

Great advances in computer technology and hardware rendering facilities enabled the rapid successes in 3D computer graphics and animation in the 1990s. Virtual worlds with increasing levels of graphical realism and complex animation have been created at relatively low cost (Badler *et al.*, 1999; Cavaza *et al.*, 1998; Millar *et al.*, 1999). Despite these remarkable achievements, virtual human animation is, relatively speaking, a weak side of the successful story, due to the fact that humans can easily perceive every unreal feature of a virtual human animation and appearance (Millar *et al.*, 1999). This therefore sets very high standards for animators to animate human behaviour.

Generally speaking, virtual character animation techniques may be categorised into two groups: motion modelling and virtual character behavioural animation.

Motion modelling mainly deals with the basic movement of virtual characters; that is, it is mainly concerned with the basic internal actions within the physical or motion constraints. It may also deal with physical body locomotion, such as collisions with the external virtual objects. Inverse kinematic techniques (Girard and Maciejewski, 1985; Boulic and Thalmann, 1992) have been studied and used to simulate complex human behaviour, such as walking and grasping. In order to use such techniques, the position of each end effector must be given; the unknown variables of internal joints may then be calculated. This type of approach is successful but there is a lack of individuality in the motion created. The use of physics-based approaches (Terzopoulos *et al.*, 1987; Brogan *et al.*, 1998) has proved to be very useful and has played a key role in creating a realistic virtual world in computer animation fields, such as video games and virtual training applications, in which the motion modelling obeys the physical laws of motion. The drawbacks of the physics-based approaches are that they have a high computational cost and that the animation generated may be predictable, which means that an animation will repeat itself exactly each time it is executed.

Motion modelling and animation emphasises physical aspects of the virtual characters' movements, whilst behavioural animation, on the other hand, deals with behavioural modelling and animation, which is a larger and more complex concept than motion animation (Hegron *et al.*, 1989; Badler *et al.*, 1999). Behavioural animation may involve the objects' characteristics and relationships developed with external objects in the virtual environment and usually related by sets of rules that govern the virtual characters' behaviour and movement at a higher level than motion modelling.

This chapter describes a new approach for behavioural simulation of virtual characters. We propose a basic two-level system. At the top level, simulation of some aspects of brain activities is performed which includes assessing, analysis, thinking, comparison and decision-making by means of AI algorithms. At the lower level, a muscle-driven virtual character model is integrated that obeys the physical laws of motion and is adjusted or commanded by the instructions or messages that come from the top level.

7.2 Related Work

Human figure locomotion synthesis and behavioural control have long been major active areas in computer graphics. Research work to date has shown impressive achievements; however, there are still many problems remaining. Efforts have been continuously made through the application of physics and dynamics, especially in terms of human behavioural animation, automatically generating human motion and goal-orientated motion synthesis.

Using forward dynamics seems less successful so far in animating the movements of active or self-actuated systems, such as living creatures, because the internal forces applied by body muscles are unknown over time. In order to avoid the difficulty and find an alternative solution to the dynamics problem, recent work has proposed methods for the use of inverse dynamics principles to simulate human locomotion (So and Badler, 1996). For a given motion, the method presented uses inverse dynamics to calculate the torques and forces required and dynamically balances the resulting walking motion to maintain the joint torque within the strength limits.

An approach was presented for animating an articulated figure motion (Gritz and Khan, 1995). The agents are dynamically controlled robots of fixed topology and geometric structure, whose behaviour is governed by a robotic controller program. This method uses the techniques of genetic programming to automatically derive a control program for the agent from the previous animation sequences and it requires the fitness metric supplied by the animator. The system developed seemed to be a little unstable and sensitive to initial conditions.

Since a numerical solution of a physics-based graphics model for rigid, articulated or dynamic models usually has a high computational cost, research work has been reported on applications of neural network for artificial agents (Grzeszczuk et al., 1998). The system was trained offline to emulate simulation through previous observations of physics-based models' animation actions. The principle of this work is similar to that of research work reported in another area (Jordan and Rumelhart, 1992).

More recently, an approach has been reported for animating articulated human figure working on slopes using parameterised motion synthesis Li and Liu (2000). Motion parameters such as slop gradient, walking speed and step length were used in the system and based on the knowledge of real-life human walking. A simulation of human locomotion was derived in which the trajectory of the centre of mass was determined according to the requirement of dynamic stability. Other interesting research recently reported was to automatically generate motion styles by learning motion patterns from a set of motion capture sequences (Brand and Hertzmann, 2000). Each sequence had a distinct choreography style. The learning model developed could identify the common choreography element, style and stylistic degrees of freedom and would then be able to reproduce an improved motion sequence for an articulated figure.

Inspired by these results, we present a different approach for human motion synthesis, using AI algorithms to control a physics-based human model whose motion is governed by the principles of direct dynamics. Our approach expands on the research above and is able to produce complex, physically plausible and aesthetically appealing motion and behavioural animation.

7.3 Human Adaptability

One of the best-known principles of learning states that we learn by experience (Bransford *et al.*, 1999). This is absolutely true of both children and adults: we learn to do what we are doing. More precisely, most of the time we learn by our direct experiences and by doing what we learn to do, by sensing and feeling, and we also learn by thinking, analysis and observation. Some other important facts that affect our learning behaviour are individual learning patterns or styles, as each person has his or her own unique way of learning and a motivation to learn. For solving a specific task, our skill and ability for that task can be developed or "evolved" during these processes of brain-doing activities or practice (Bransford *et al.*, 1999). This concept can be summarised as shown in Figure 7.1 and will be used to form a basis for our virtual character behavioural simulation model.

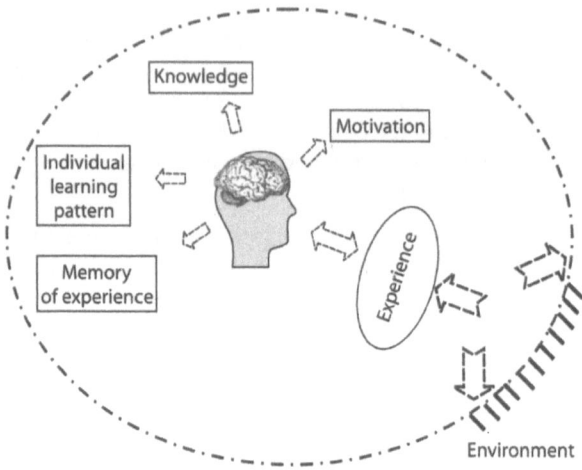

Figure 7.1 The role of the human brain.

7.4 The Concept of Behavioural Simulation

Although the human brain is not a computer at all (it is virtually impossible to use a computer to do all the things that a human brain is usually capable of doing), we can certainly simulate some of the human brain's activities. The number of brain activities we could use a computer to simulate today has dramatically increased and this tendency appears to continue.

The concept of our simulation system can be summarised as shown in Figure 7.2. We propose a general virtual character behavioural animation system that is actually of a goal-orientated two-level structure. At the top level, a behavioural control unit (artificial brain unit) consists of a number of key components including motivation, behavioural planning, a knowledge database,

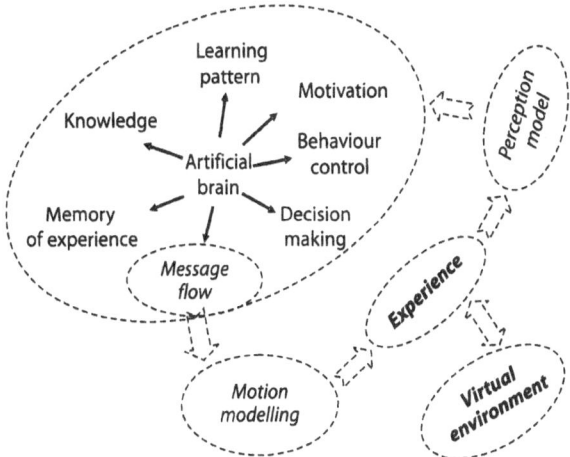

Figure 7.2 A general behavioural simulation model.

memory of past experience and decision making. The exact structure of the components may vary from one to another, depending on the specific behavioural simulation application. In this system, the control unit commands all the behaviours and reactions of the virtual character for a specific task. It perceives stimuli from the virtual world, assesses a virtual character's own behavioural performance, searches for an optimal solution for the task and passes the message to the action model that is at the bottom level of the system.

7.5 Learning in Virtual Environments

To achieve our goal, there are two fundamental frame tasks which must be done: human locomotion synthesis and behavioural control.

The principles of forward dynamics are adapted and used for human model motion synthesis. However, forward dynamics alone cannot determine the actual model locomotion. It is because of this that we cannot use forward dynamics to predict how the human body will walk, nor to predict how the walk will change in response to an external force acting on the model if we are not given the required body joint force or torques.

The behavioural control unit is therefore proposed and developed for planning, learning and determining the muscles' activities and body joints' force patterns required for a specific locomotion task. A good experience of muscle activities or patterns will be remembered or stored in a memory database in the control unit of the simulation system for future use. This is exactly what our human brain and body muscles are doing in reality and the approach proposed in our work is therefore a closer simulation of human motion and behaviour compared with other approaches.

In this section, we will discuss the methods and algorithms that have been implemented for the muscle-based human model and its behavioural control.

7.5.1 Physics-based Human Model

We use direct forward dynamics to determine and predict the outcome of a human figure's locomotion, which is derived from the muscle forces. The initial pattern of muscular forces for a particular task will come from the experience memory database. The initial pattern of muscular forces may be just one type of time-domain sequence or a combination of several types of time-domain sequence, which depends on how complex the task is.

A physical network is therefore developed as the low level in our simulation system, which is an integrated model that deals with motion modelling and synthesis, action implementation, and also performance feedback. The core of this level is a muscle-driven virtual character model in which we use muscle-contracting actions to control a virtual character's movement. Our goal is to simulate human movement in a more realistic way, as humans utilise their muscles in many ways in order to walk or run. Producing physically correct motion is not equivalent to producing natural-looking motion, and that is why the virtual human needs to learn how to use its muscles. A typical example of the concept is shown in Figure 7.3, which illustrates that a knee joint utilises two muscle pairs to fully control its movement.

In order to be adapted for accessing messages from the behavioural control units effectively, one has to find a complete set of parameters that could fully determine the human body locomotion. We proposed a low-level parameter space for describing the characteristics of the muscular forces and for governing body locomotion using time-domain muscular variables rather than high-level parameters such as walking speed, step length and walking frequency (Li and Liu, 2000), which can be obtained directly through motion capture methods (So and Badler, 1996; Brand and Hertzmann, 2000). In order to control and optimise the body locomotion, the fewer the parameters used, the higher the implementation efficiency, because it reduces the size of the parameter search space. Muscle units can be considered as energy engines in the model. The parameters used in our model currently consist of a number of key variables which are related to the muscular forces in terms of amplitude index, shape function, energy limits, energy burst type, muscular state and action

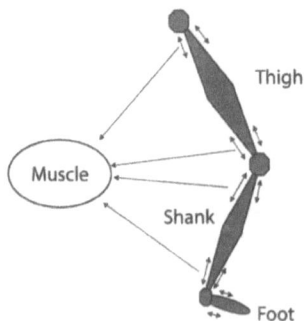

Figure 7.3 An illustration of the muscle driven model.

period, plus kinematic constrains which guarantee that the body locomotion synthesised is physically plausible.

If $\mathbf{p}_i$ denotes the ith joint location vector, $\boldsymbol{\theta}_i$ its angle vector, and $\boldsymbol{\tau}_{i,j}$ the torque vector located by the muscular force vectors nearby, the state of body locomotion can be determined by the equations

$$\mathbf{f}_{i-1,i} = m_i \frac{\mathrm{d}^2 \mathbf{p}_i}{\mathrm{d}t^2} + \mathbf{f}_{i-1,i} - m_i \mathbf{g}$$

$$\boldsymbol{\tau}_{i-1,i} = \boldsymbol{\tau}_{i-1,i} + \mathbf{c}_{i-1,i} \times \mathbf{f}_{i-1,i} - \mathbf{c}_{i,i} \times \mathbf{f}_{i,i+1} + \mathbf{I}_i \frac{\mathrm{d}^2 \boldsymbol{\theta}_i}{\mathrm{d}t^2} + \frac{\mathrm{d}\boldsymbol{\theta}_i}{\mathrm{d}t} \times \mathbf{I}_i \frac{\mathrm{d}\boldsymbol{\theta}_i}{\mathrm{d}t}$$

where m_i is the mass of the body part considered, $\mathbf{f}_{i,j}$ is the force applied from an adjacent body part or external object and $\mathbf{I}_i$ is the centroidal inertia tensor of the body part considered. Since the force $\mathbf{f}_{i,j}$ is related to the adjacent body parts or external objects, an efficient method is to use a recursive algorithm to solve the equations.

We use a numerical algorithm to find the solution of the equations above. In general, using $\mathbf{p}_i$ to denote the position vector i from the current state to compute the state at the next time step, we have the formula

$$\mathbf{p}(t + \Delta t) = \psi \left(\Delta t, \mathbf{p}(t), \frac{\mathrm{d}\mathbf{p}(t)}{\mathrm{d}t}, \frac{\mathrm{d}^2 \mathbf{p}(t)}{\mathrm{d}t^2}, \Delta \mathbf{p}(t), \Delta \frac{\mathrm{d}\mathbf{p}(t)}{\mathrm{d}t}, \Delta \frac{\mathrm{d}^2 \mathbf{p}(t)}{\mathrm{d}t^2} \right)$$

7.5.2 AI Algorithms and Behavioural Control

Physics-based behavioural control has been a challenging task in human behavioural modelling. To simulate a complex human behaviour, what we need to do is to search for and find a solution in terms of a set of parameter values or functions which can be regarded as the input to the physical system. The whole process is actually an optimisation. Physics-based human models are usually represented as non-linear systems and their behavioural control seems to be formidable, or with very high computational cost, especially for complex human behavioural simulation. Therefore conventional methods have been comparatively successful. Here we propose a novel approach which utilises GA (Genetic Algorithms; Goldberg, 1989) to implement the optimisation to determine the best set of parameter values for the system input.

Searching and Optimisation

The principle of genetic algorithms is used in our work for AI-based parameter optimisation. Genetic algorithms are search procedures based on the mechanics of natural selection and genetics. GA-based methods are quite different from conventional methods in many ways. They are robust, problem-independent, parallel algorithms and have good overall performance as a whole. However, it may be slow in convergence speed. In a GA implementation, each solution is well defined in the state variable space and treated as an individual or a child. A collection of all current individuals forms a population, and the individuals with fitness rates, together with some individuals selected randomly from the current population, will be responsible for producing the next generation of the population.

In order to implement a GA-based technique, we have to derive a state variable space from the physics-based model considered, which must be suitable for a GA-based algorithm to act. In the current work, the goal is to simulate human working and jumping. The state variable space is represented in the form of a set of parameters or state attributes. For a typical body joint, a state parameter set for a joint is shown below:

```
State Attributes{
P1: Joint index;
P2: Amplitude index;
P3: Motion and Location Variables; (parameters related to posi-
tion, velocity acceleration of the joint)
P4: Energy burst type;
P5: Muscle force shape function;
P6: Muscle State;
P7: Kinematics constrains;
P8: Energy Limit; (Maximum values etc.)
P9: Action Period;
...
}
```

In the implementation, as shown in Figure 7.4, those parameters are represented as real numbers and constructed in the form of a chromosome string, which is therefore suitable for a GA-based method to manipulate.

Conventional genetic algorithms usually need a searching space with a very large population, which may be over hundreds or thousands of individuals and the initial population is created randomly. This is not a suitable situation for simulating complex human behaviour. In our work, we let the model start its attempt from a much smaller space initialised by well planned or experienced individuals, or seeds, which are initially stored in the memory database of our simulation system. After the virtual model has tried all the potential individuals in the population, the individuals' performances will be analysed and those individuals with a good performance record will be chosen, together with some randomly generated ones, to be responsible for creating a new generation of possible individuals or solutions. The criteria that are set could be based on the performance rate, including an index that indicates a measure of how long it

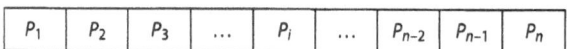

Figure 7.4 Chromosome representation of an optimisation problem.

takes to reach its target. The search process is completed once the criteria are met. In this way, the required searching space is reduced and the efficiency of reaching an optimised solution is increased significantly. This process is a simulation of reality.

Genetic Operators

To produce a new generation from the chosen individuals, three basic genetic operators are considered and used: Cross-Over, Mutation and Elitism. Cross-Over is performed by taking two parent individuals to produce two children through information exchange between them. Mutation produces a child by creeping any value up or down a small and random amount. Elitism reproduces itself as a way to keep the best set of parameter values in the succeeding parameter sets.

Performance Evaluation

The resultant performance may be assessed by a set of rules and/or directly by the goal to which the virtual character is intended to achieve. Once the virtual character performs an action, the resulting performance rate is fed back for comparison and analysis. Using the results obtained, the system will then decide if there is a need to further update the parameter's structural pattern for a better solution. In this way the virtual character is being trained to know how to do it the next time that this specific task appears.

7.6 Implementation

Figure 7.5 shows an example of a simulation sequence of an articulated human animation, in which we simulate one of the complex virtual human's behaviours: learning to jump. As shown in the figure, this is not a simple jump but a combination of running and jumping. The target of the action is to reach a certain height. In order to complete this action, the human figure had to go through a number of complex manoeuvres using its muscles, and at the same time keep itself in good dynamic balance.

Figure 7.6 shows a simulation sequence of jumping in a virtual 3D world. The virtual man is trying to grab an apple from an apple tree, but it may be not an easy task for him. He might not be able to jump to that height at his first jump. In order to reach his goal, he has to jump as high as he can by learning through a number of attempts of running and jumping. Let us suppose that he could do it physically; the task is only a matter of skill, and all he needed was to develop his skill by learning from his experience of jumping until his skill developed enough for him to grab the apple.

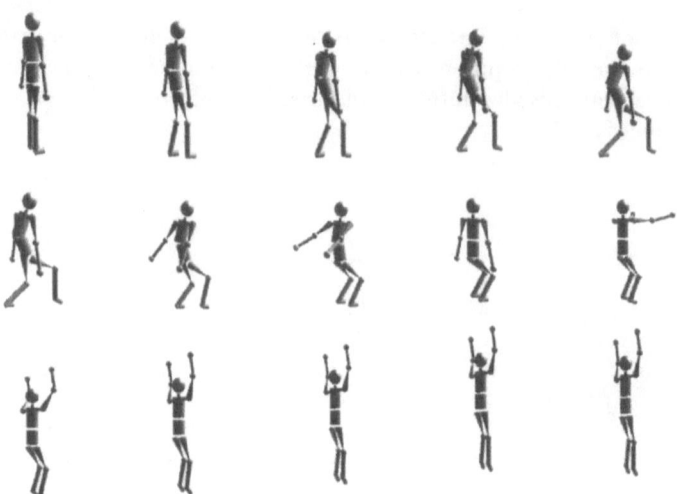

Figure 7.5 A sequence of an articulated human jumping simulation.

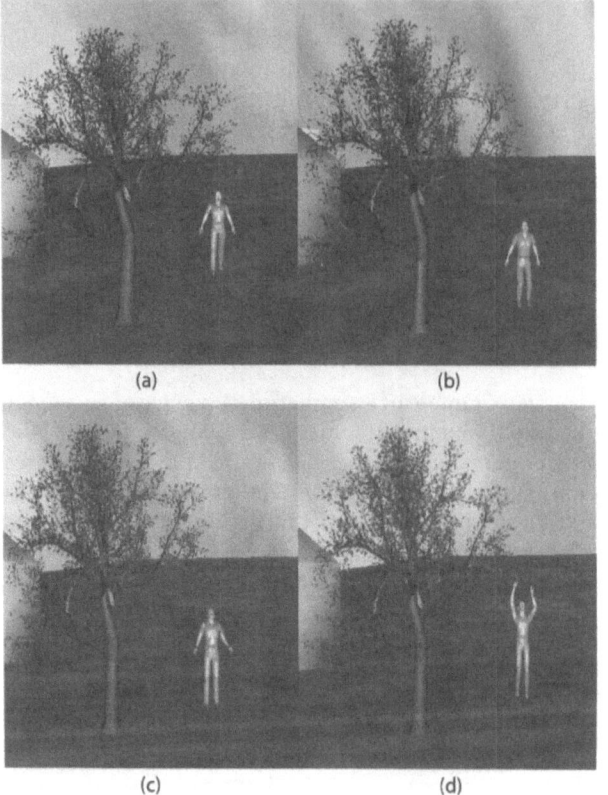

Figure 7.6 A sequence of human figure jumping in a virtual 3D world.

7.7 Conclusion

Satisfactory simulations may be involved in many disciplines, including computer science, mechanics, physics and biology. In order to model and animate virtual autonomous characters realistically, there are many techniques that have been reported, most of which have been focused on the use of a set of rules of behaviour to control the animated creatures. A framework for simulating complex virtual human behaviour has been developed in our research, although the current work is still at early stage. Our proposal also needs the rules of behaviour, but one step further beyond that. Generally speaking, most of the current approaches are dealing with tasks, such as "a move" (motion modelling) and "to move to a place targeted" (behavioural animation). Our work is to study not only those mentioned above but also "to learn how to do it". The concept of an artificial brain is presented, and it therefore provides a way in which autonomous virtual characters will be able to collect and analyse information through their experiences in a virtual world. The experiment shows that this approach is suitable for simulating real humans' activities, thanks to the muscle-driven model developed, together with other parts of the system that make it possible to implement the whole ideal. Our future work will continue to improve the simulation system in order to simulate more complex human learning behaviours, such as learning to dance, and to look at applications in video games and virtual training systems.

References

Thomas, F. (1984) Can class Disney animation be duplicated on the computer? *Computer Pictures*, 2(4), 20–26.

Badler, N. I., Bindiganavale, R., Bourne, J., Allbeck, J., Shi, J. and Palmer, M. (1999) Real time virtual humans. In *Proceedings of International Conference on Digital Media Futures*, British Computer Society, Bradford.

Cavaza, M., Earnshaw, R., Magnenat-Thalmann, N. and Thalmann, D. (1998) Motion Control of Virtual Humans. *IEEE Computer Graphics and Applications*, 18(5), 24–31.

Millar, R. J., Hanna, J. R. P. and Kealy, S. M. (1999) A review of behavioural animation. *Computers and Graphics*, 23, 127–143.

Girard, M. and Maceijewki, A. A. (1985) Computational modelling for the computer animation of legged figures. In *SIGGRAH'85, Computer Graphics*, 19(3), 263–270.

Boulic, R. and Thalmann, D. (1992) Combined direct and inverse kinematic control for articulated figure motion editing. *Computer Graphics Forum*, 11(4), 189–202.

Terzopoulos, D., Platt, J., Barr, A. and Fleischer, K. (1987) Elastically deformable models. *SIGGRAPH'87 Computer Graphics*, 21(4), 205–214.

Brogan, D. C., Metoyer, M. P. and Hodgins, J. K. (1998) Dynamically simulated characters in virtual environments. *IEE Computer Graphics and Applications*, 18(5), 25–34.

Hegron, G., Palamidese, P. and Thalmann, D. (1989) Motion control in animation, simulation and visualisation. *Computer Graphics Forum*, 347–352.

Badler, N. I., Bindiganavale, R., Bourne, J., Allbeck, J., Shi, J. and Palmer, M. (1999) Real time virtual humans. *Proceedings of International Conference on Digital Media Futures*, British Computer Society, Bradford.

So, H. and Badler, N. I. (1996) Animating human locomotion with inverse dynamics. *IEEE Computer Graphics and Applications*, March, 50–59.

Gritz, L. and Khan, J. K. (1995) Generic programming for articulated figure motion. *The Journal of Visualisation and Computer Animation*, 6, 129–142.

Grzeszczuk, R., Terzopoulos, D. and Hinton, G. (1998) NeuroAnimator: fast neural network emulation and control of physics-based models. *Computer Graphics (SIGGRAPH '98 Proceedings)*, Orlando, FL, July, pp. 9–20.

Jordan, M. I. and Rumelhart, D. E. (1992) Supervised learning with a distal teacher. *Cognitive Science*, 16, 307–354.

Li, L. and Liu, X. (2000) Simulating human walking on special terrain: up and down slopes. *Computers and Graphics*, 24, 453–463.

Brand, M. and Hertzmann, A. (2000) Style machines. *Computer Graphics (SIGGRAPH 2000 Proceedings)*, New Orleans, July, pp. 183–192.

Bransford, J.D., Brown, A. L. and Cocking, R. R. (1999) *Brain, Mind, Experience, and School*. Washington, DC, National Academy Press.

Goldberg, D. E. (1989) Genetic algorithms in search. In *Optimisation, and Machine Learning*. Reading, Addison-Wesley.

8

The Development of an Intelligent Virtual Environment for Training

Ian Palmer, Donna Robey, Nic Chilton, Jurgen Dabeedin, Patrick Ingham and Simon Bramble

Abstract

Crime scene investigation generally involves a substantial amount of time at the scene itself. This is often inconvenient, and there is not enough time or resources to satisfactorily complete the investigation. This may be due to a limitation on the time that the scene can be kept sealed or due to the geographical disparities or schedules of colleagues. Computer graphics has become more common in crime scene investigation, for example in the 3D representation of objects (Little *et al.*, 2000). With computers continually increasing in power, new tools are available that allow the operator to interact, in real time, with a computer-generated virtual environment. This chapter explores the potential use of such real-time virtual environments as an additional tool in crime scene investigation. It discusses an environment that integrates a virtual environment with an intelligent system.

8.1 Virtual Reality Technology and Crime Scene Investigation

Virtual reality (VR) technology can be used in crime scene investigation in a number of ways. Previous research in the field has produced this (non-exhaustive) list (Howard *et al.*, 2000):

- Data analysis
- Witness statement evaluation
- Witness assistance
- Route visualisation
- Briefing officers
- Hypothesis evaluation

- Training
- Lighting evaluation
- Security planning
- Crime prevention
- Courtroom uses

The reconstruction of the crime scene and artefacts has been a focus of much of the previous work (http:/www.esat.kuleuven.ac.be/~konijn/improofs.html). The Forensic Science Service (FSS) is a partner in a project investigating computer vision techniques for analysing images arising in forensic applications funded by the European Union (ESPRIT project No. 23515). Substantial research has been carried out into obtaining 3D information from multiple images and constructing a 3D virtual reality model of the scene as part of the Image Processing Operations for Forensic Science (IMPROOFS) project (http:/ www.robots.ox.ac.uk/~improofs/; Murta *et al.*, 1998). The REVEAL project (http://aig.cs.man.ac.uk/research/reveal/) is also concerned with this application, with the emphasis on the accurate reconstruction of the crime scene itself.

Modern techniques must be used for the storage, visualisation and manipulation of evidence that are vital if investigative resources are to be used efficiently. Future crime investigation units are likely to be small but highly technological, allowing the acquisition and cataloguing of many data types quickly and efficiently (Baldwin, 1999). As the complexity and quantity of data increases, so does the difficulty in analysing and categorising it. VR technology offers an attractive interface to manipulate this data.

Three areas are seen for possible application of VR technology:

1. *Intelligent agents*

    In this, an expert system is integrated into the virtual environment. This stores specialist knowledge gained by experts from previous investigation work and so provides online guidance at key points relating to the current crime scene. Examples would be advice about the correct procedure for various evidence types, notifying the user when particular scene content may be of interest to officers from other disciplines (e.g. bio, DNA, firearms) and providing links to other previously processed scenes where similar evidence may have been found. The user could annotate the scene content and then the system could update the knowledge base accordingly. This dynamic update feature could prove crucial in the investigation of ongoing burglaries and car crimes.

    The same knowledge base can form the basis of a training system. The agent would then perform the role of tutor, guiding the user and offering advice. For this application, the system could use synthetic scenes produced specifically for the training application or they could be based on real crime scene data. This particular application will form the basis of the proposed system, as it allows the complexity and the content of the scene to be completely controlled.

2. *Multi-user collaborations and scenario testing*

Once a crime scene has been reconstructed in VR, several users can then share the environment simultaneously. This allows different experts, probably located at different locations, to share information and discuss the case whilst in a reconstruction of it. It would also be possible to play out different scenarios interactively with all users sharing their knowledge to optimise the recovery of evidence from crime scenes. Scene examiners, scientists and investigating officers could share information from different scenes and/or laboratory and make use of the intelligent agent for information and guidance. Sharing information in this way could make the whole forensic process more efficient, reliable and timely.

3. *Visualisation of complex data*

There can be a vast amount of data collected from a crime scene, many of it involving complex technical data. It is possible to use a 3D representation of this data to make explicit the spatial relationships and connections between different types of evidence. This can aid interpretation and analysis. The use of hyperlinks between evidence representations and other information such as photographs, technical notes, laboratory reports, etc. comes under this heading.

This chapter describes a system that exhibits two features discussed in this list: the incorporation of an intelligent agent and the support of multi-user collaboration.

VEs that incorporate intelligent features can be significantly easier to use and can offer improvements over standard VEs (ECAI, 1998; Applied Artificial Intelligence, 2000). The intelligent features can provide an easier to use and richer environment by supporting higher levels of information and interaction. This can lead to significant improvements in efficiency for users of the environment.

As the system was conceived from the outset as a multi-user system, the choice of software and hardware was important. The system needed to run on relatively modest hardware and to be capable of easily supporting multiple users.

8.2 The Prototype Toolset

The prototype is based around two sub-systems: the virtual environment and the knowledge database. These communicate via a socket-based connection so the two elements can run on different machines in different locations. This allows the use of a single centralised knowledge database with multiple VEs representing different crime scenes. It would also be possible to distribute the database across multiple machines, i.e. creating a distributed multi-database system.

8.2.1 System Architecture

Figure 8.1 shows the system architecture. The VE clients and server run on Windows NT systems. The system is relatively inexpensive to allow easy distribution and use in a multi-user configuration. More advanced hardware is being

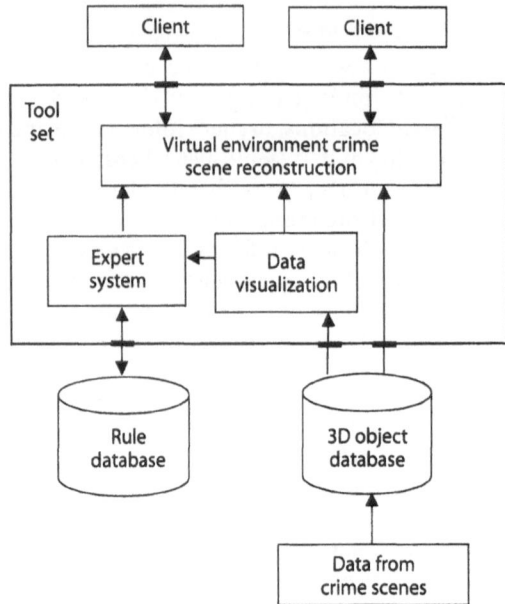

Figure 8.1 The system architecture.

consideration for future systems to support immersive and semi-immersive. The central tool set consists of the reconstruction of the scene, the expert system and the data visualiser. In the current system only the reconstruction and the expert system are implemented, the visualiser being left for later development.

8.2.2 System Software

The system was initially prototyped in DIVE (http://www.sics.se/dive/), VRML (http://www.Web3d.org/technicalinfo/specifications/vrml97/index.htm) and Unreal™ (http://www.unreal.com/). Screenshots of these three implementations are shown in Figure 8.2: DIVE at the top left, VRML in the CosmoWorlds browser, top right, and Unreal at the bottom of the figure. Unreal was chosen over the other two solutions because of the following features:

1. Multiple users can be supported over low-bandwidth networks (e.g. standard modems).
2. The graphics engine supports high-quality interactive graphics.
3. It is easy to introduce new features through the built-in scripting language (UnrealScript) and its own C++ API.
4. The client software (the Unreal game) is readily available and will run on standard PC hardware.

There are already some research projects based on the Unreal games engine (Applied Artificial Intelligence, 2000; http://www.unrealengine.com/; DeLeon, 1999;

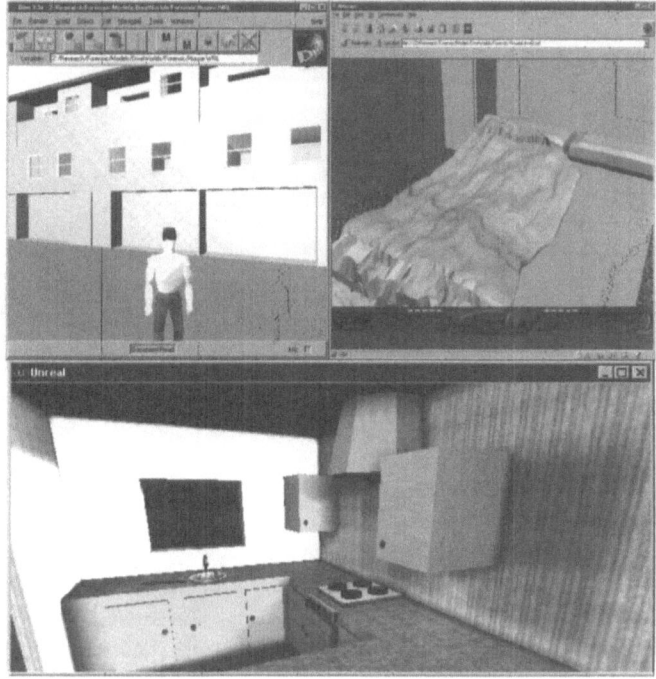

Figure 8.2 The test scenario running in three different environments.

http://www.vrndproject.com/; http://www.unrealty.net/). These have proved its suitability for use in high-quality virtual environment applications. The game engine is supported by a dedicated application to construct content, UnrealEd. For some of the more complex modelling, 3DStudio Max™ has been used and the models then imported into UnrealEd.

The expert system needs to be incorporated in the Unreal environment, and there are several choices for this. In the first prototype the AI was implemented in the UnrealScript language. This has the advantage that it is fully integrated into the 3D environment, i.e. it can access and modify the 3D data directly. UnrealScript support for simple AI through features such as state-based execution, but is ultimately limiting when developing large scale AI systems. Figure 8.3 shows a screenshot of the UnrealScript-based system. This system proved suitable for the relatively limited prototype, but since the ultimate aim is to produce a fully fledged expert system, it was decided to seek a more flexible and extendible solution.

To develop the system further, a way of not only developing a more fully featured AI system but also of integrating this into the VE is needed. The Java Expert System Shell (JESS) has been chosen as the basis for this (http://herzberg.ca.sandia.gov/jess/). This has been used in a number of experimental systems and was chosen because:

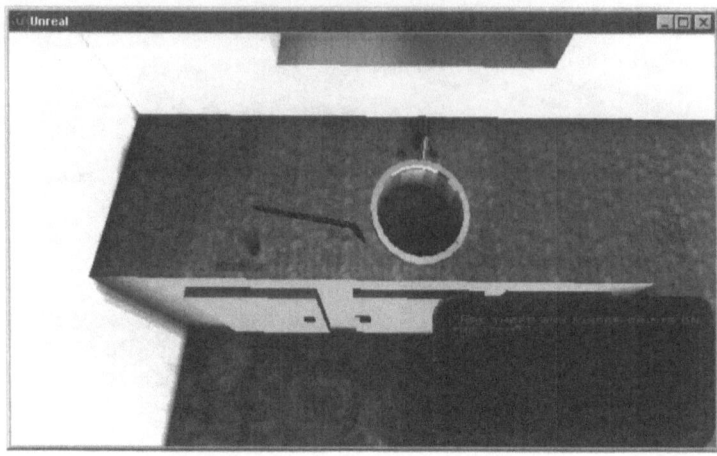

Figure 8.3 The Unreal environment using UnrealScript for the AI.

1. It supports a library of functions for logic and rule-based system development.
2. It provides easy access to the standard Java features, such as the networking libraries.
3. Being Java-based it is available for multiple platforms.

These features have allowed the prototype to be developed quickly and with minimal effort. JESS can operate in two ways: as a command line-based

Figure 8.4 The JESS expert system shell running in a separate window.

Figure 8.5 The integrated system interface.

interactive shell and by writing dedicated Java programs that use the JESS class libraries. Initial development of the expert system focused on the shell so that it could be tested quickly. Because of this, initially the expert system was not integrated into the environment. This meant that to use the VE and the expert system it was necessary to use two windows: one graphical window with the VE and one command line with the expert system shell running (see Figure 8.4). This allows the rapid development and testing of the two sub-systems, but is obviously far from ideal since the user has to constantly switch between the two windows while trying to navigate the environment and absorb the information. The final system uses a single interface, with the JESS program communicating with the user through the Unreal interface as shown in Figure 8.5.

8.3 The Scenario Used to Develop and Test the System

To effectively test and develop the system, a scenario is needed. This could either be based on a real-life crime scene or dedicated one could be produced. The latter was chosen, as it allows complete control the content of the scene, creating elements as we deem appropriate. There are also no legal implications in using a fictional scenario.

The events in the fictional crime occur as follows:

1. A burglar finds the garage door of the victim's house unlocked. The door pivots open using central handle operation.
2. Once the burglar is inside he shuts the garage door behind him so that it seems locked from outside.
3. The burglar then forces the internal garage door open with a crowbar brought to the crime scene.
4. The burglar has now gained access to the downstairs hallway. Carrying the crowbar, he turns right and climbs the stairs to the top floor. He then proceeds to take valuables from the bedroom.

5. Meanwhile the victim comes home earlier than expected. She enters through the front door and climbs the stairs to the first floor, where she enters the kitchen and places her shopping bags on the floor. She is unaware of the break-in.
6. The burglar hears noise and tries to make a quiet exit, but is seen by the victim as he gets to the first floor landing.
7. The victim reacts by picking up a heavy glass and throwing it towards the burglar's head.
8. The glass hits the target, which then starts to bleed heavily.
9. The burglar lifts his hand to the wound, staggers, and with the same hand touches the kitchen doorframe to regain balance.
10. In a rage the burglar goes for the victim and strikes her with the crowbar he has been holding.
11. The crowbar causes an extensive wound to the back of the skull. The victim slumps to the floor and her wound starts to bleed heavily, dripping onto the floor.
12. The burglar, who is standing over her, realises that she is badly hurt.
13. Blood from the burglar's head wound drops on to the victim's body.
14. Realising he has blood on his hands and face, the burglar goes to the sink, puts the crowbar down on the worktop, and washes his hands and bleeding eyebrow. His efforts are in vain, however, as his wound continues to bleed.
15. In a panic, the burglar runs to the hallway, forgetting his crowbar, stepping in his and the woman's blood pools on the kitchen floor.
16. The burglar grabs his bag of stolen items from the hallway and runs downstairs, exiting through the front door, which he leaves slightly ajar.

This gives us a very varied selection of evidence types whilst keeping within the realms of realism. Evidence would include:

1. Latent marks on and around the handle of the garage door.
2. Latent marks and tool marks on the internal garage door.
3. Latent marks and blood marks on the crowbar.
4. Latent marks in the upstairs bedroom.
5. Blood and latent marks on the glass thrown at the burglar.
6. Blood marks on the kitchen doorframe, left by hand.
7. Latent marks on the kitchen taps.
8. Blood residue in and around the kitchen sink, including inside the U-bend.
9. Pool of blood from the victim's head on the floor.
10. Blood drops on the victim's clothing from the burglar's wound.
11. Blood drops on the kitchen and hallway floors from the burglar's wound.
12. Footprints made by the burglar, from stepping in blood on the kitchen and hallway floors.
13. Blood marks on the internal face of the front door.
14. Blood marks on the internal handle of the front door.
15. Woman's body on the floor.

The layout of the building in which the fictional crime occurs is shown in Figure 8.6. The position and type of evidence is also shown in the diagram.

For the prototype, the contents of the rules in the expert system have been based only on the evidence types and features present in this test scenario. Obviously in the final system this would in fact be based on knowledge acquired from real experts and would be applicable to multiple scenes. Although this is obviously limiting, the current system does allow the testing of different versions of the scenario (i.e. with or without certain pieces of evidence) to allow evaluation of the effectiveness of the VE.

As the user moves around the environment, prompts are triggered from the expert system to the user. Many of these are questions which require "yes/no" type answers, while others identify types of evidence or offer directions to the user. The rules are stored in a simple text file at present, although it is expected that a relational database will be used for the knowledge base in the next

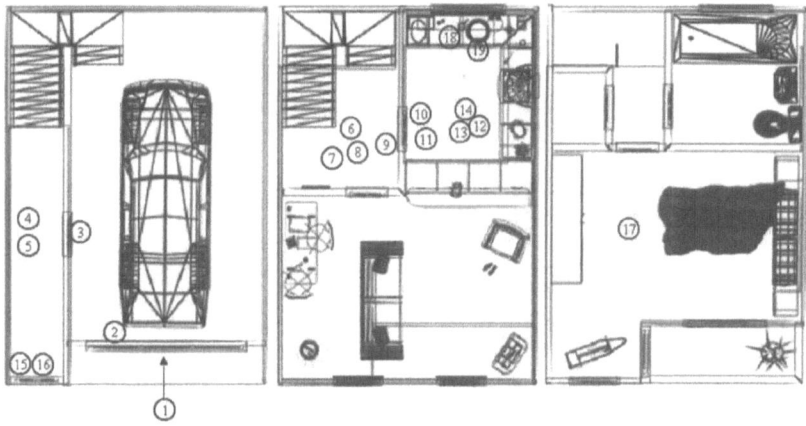

1. Latent marks on and around handle of garage door.
2. Latent marks on inside of external garage door.
3. Latent marks and tool marks on internal garage door.
4. Footprints, in blood, leading to front door.
5. Blood drops on carpet.
6. Footprints, in blood, leading to downstairs hallway.
7. Blood and latent marks on upturned glass.
8. Blood drops on carpet.
9. Blood marks on kitchen doorframe.
10. Blood pools on floor.
11. Footprints, in blood, on floor
12. Woman's body on floor.
13. Pool of blood by woman's head
14. Blood drops on woman's clothing.
15. Blood marks on internal side of front door.
16. Blood marks on internal handle of front door.
17. Latent marks in bedroom.
18. Latent marks and blood marks on crowbar.
19. Blood residue in and around sink.

Figure 8.6 The scene layout with location and type of evidence.

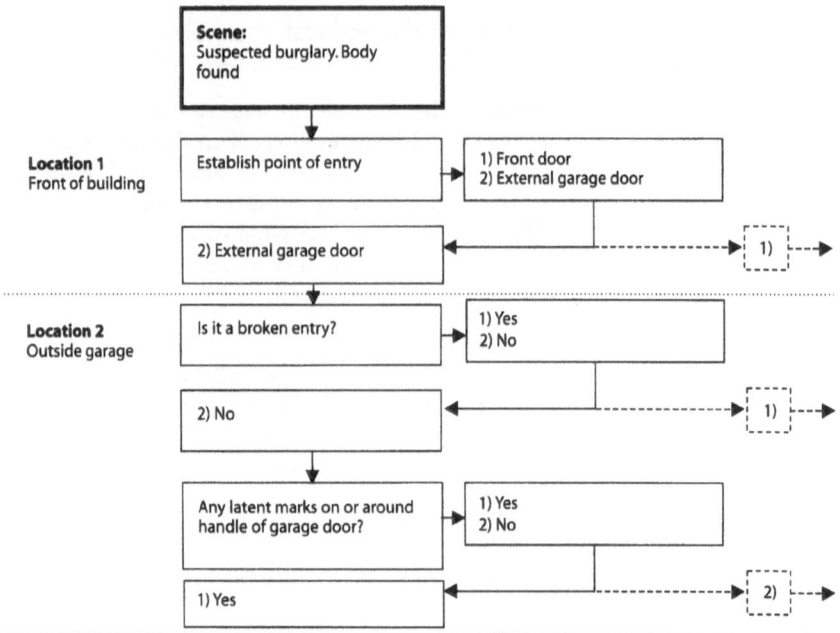

Figure 8.7 Extract from a typical user–expert system interaction.

iteration. This is parsed by the expert system at initialisation. A typical user–expert system interaction is shown in Figure 8.7.

8.4 Conclusions and Future Work

The prototype system demonstrated that the creation of intelligent VEs for crime scene reconstruction is viable using existing technology. The expert system developed provides a simple textual interface that is integrated into the VE. This offers an unnatural interaction and can prove distracting to the user. It is proposed that in the next version the expert will have a virtual representation as an avatar. This will appear when the system is called upon for advice, but will otherwise not be present in the system.

The rules present in the expert system are limited to those relevant to the test scenario. This needs to be developed to incorporate enough data to support more scenarios. This is obviously a major task in itself and will form a complete project that could be used in a standalone way without the VE. This can then operate in one of two ways: it could be called up and questioned as needed (the "mentor" mode), or it could act as a proactive guide providing more "hands on" help (the "trainer" mode).

It is also desirable to develop a speech-based interface to the system. This will allow more natural communication between the user and the system, and so would also allow users to concentrate on navigation and interaction with the

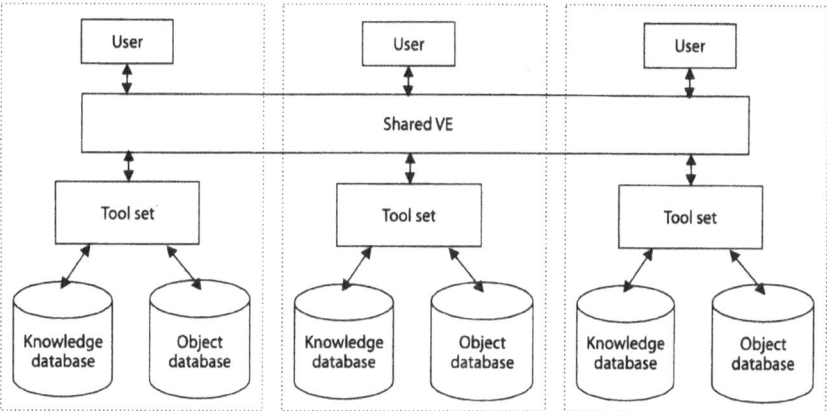

Figure 8.8 Proposed multi-database system.

environment, as they would not have to stop and type in responses. This will build on previous work in the area incorporating an efficient NLP-based inter-action with the expert system (Cavazza and Palmer, 1999, 2000a).

The prototype system utilises a single knowledge base and a single graphics database. Future systems will provide support for multiple databases, each using its own toolset to feed data into the shared VE. The proposed architecture of the system is shown in Figure 8.8. This would then allow investigation offices to develop their local own knowledge bases, but to share the expertise with other offices.

Further development is planned to provide action recognition within the VE as has already been developed for other VEs (Cavazza and Palmer, 2000b). This would then be used to provide higher-level data for processing by the expert system, which would then advise action appropriate to a given situation without a direct request from the user. It could also provide performance assessment of the user by logging and analysing the user's actions in the training application.

Acknowledgements

Thanks go to Hamdan Amin for his help in developing the DIVE version of the prototype and to Joanne Morgan for collating the forensic information and helping to develop the test scenario.

References

Little, C., Small, D., Peters, R. and Rigdon, J. (2000)
Forensic 3D scene reconstruction. *28th AIPR Workshop: 3D Visualisation for Data Exploration and Decision Making, Proceedings of SPIE*, **3905**, 67–73.
Howard, T. L. J., Gibson, S. and Murta, A. D. (2000) Virtual environments for scene of crime reconstruction and analysis. *Proceedings of SPIE/IS&T*, **3960**, 41–48.

Murta, A. D., Gibson, S., Howard, T. L. J., Hubbold, R. J. and West, A. J. (1998) Modelling and rendering for scene of crime reconstruction: a case study. In *Proceedings Eurographics UK*, Leeds, March, pp. 169–173.

Baldwin, H. B. (1999) Crime scene units: a look to the future. *Investigation and Forensic Science Technologies, Proceedings of SPIE*, 3576, 47–50.

ECAI (1998) *Proceedings of ECAI 98 Workshop on Intelligent Virtual Environments*, Brighton.

Applied Artificial Intelligence (2000) *Applied Artificial Intelligence: Special Issue on Intelligent Virtual Environments*, 14(1), January.

DeLeon, V. J. (1999) VRND: Notre-Dame Cathedral: a globally accessible multi-user real-time virtual reconstruction. *Proceedings of VSMM99*, Dundee, September.

Cavazza, M. and Palmer, I. J. (1999) Natural language control of interactive 3d animation and computer games. *Virtual Systems and Multimedia '99*, Aberdeen, September.

Cavazza, M. and Palmer, I. J. (2000a) High-level interpretation in virtual environments. *Applied Artificial Intelligence Special Issue on Intelligent Virtual Environments*, 14(1), 125–144.

Cavazza, M. and Palmer, I. J. (2000b) Natural language control and paradigms of interactivity. In *Proceedings of the AAAI Symposium on Artificial Intelligence and Interactive Entertainment*, 20–22 March, Stanford University, Palo Alto, California.

9

An Experimental Study of the Effect of Presence in Collaborative Virtual Environments

M. Gerhard, D. J. Moore and D. J. Hobbs

Abstract

This chapter explores one aspect of interaction in virtual environments, namely the degree of "presence" experienced by participants in relation to the avatar style used to represent them. A prototype virtual art gallery was created as a vehicle for conducting a series of online experiments designed to measure and compare the effects of different forms of avatar on presence. The choice of experimental procedure, together with analysis and interpretation of the results, are presented.

9.1 Introduction and Rationale

The user–computer interface has undergone many changes, from the textual interfaces of the 1970s to graphical interfaces in the 1980s, from incorporation of multimedia elements in the 1990s to current work in portraying sophisticated virtual environments. At the same time, systems have evolved from being single-user-oriented to sophisticated collaborative multi-user systems. As a result, traditional HCI (Human–Computer Interaction) guidelines are limited in their applicability for the design of such systems and are consequently unlikely to address the full range of aspects now inherent in these virtual environments.

Further, there are currently no evaluation methods specific to CVEs, and differences between virtual environments and conventional interfaces are not fully understood (Kaur *et al.*, 1998). Consequently, defining an evaluation methodology is complex, but it can at least be argued that the specific diet of evaluation techniques needed depends on the characteristics of the system to be evaluated as well as on the purpose of the evaluation itself (Tromp and Benford, 1996). Whilst the experimental design and the evaluation of the experiment described in this chapter are founded on general HCI usability principles, it is argued that the approach adopted nevertheless forms a unique approach to CVE usability evaluation.

The factor that allegedly distinguishes CVE technology is the sense of immediacy and control created by presence: the feeling of "being there" (Psotka, 1995). It is this factor, therefore, which will be addressed via the evaluation approach just discussed. The term *presence* as used in this context is understood as the possible result of the process of cognitive immersion, and is not to be confused with *virtual presence*, which is simply the deployment of avatars within CVEs. *Presence* may be felt in varying degrees (including no feeling of presence) and may or may not be caused by the use of avatars.

An avatar is "*the representation of a user's identity within a multi-user computer environment*" (Gerhard and Moore, 1998). In other words, an avatar is a proxy for the purposes of simplifying and facilitating the process of inter-human communication in a virtual world. The use of avatars potentially entails several useful properties within a virtual environment, in particular identity, presence, subordination, authority and social facilitation. Avatars may provide a way for other users to better understand the actual or assumed persona of the underlying user. They may help establish a feeling of presence within a multi-user virtual environment. They may imply subordination, being under the direct control of the user, without significant control over their own actions and internal state. Avatars may also facilitate social encounters in the virtual world and may imply to others that they are acting with the authority of the underlying user.

Based on a review of existing CVE applications and literature, a theoretical framework for understanding the relevance of user embodiments within a CVE for education was expounded by Gerhard and Moore (1998). They argue that presence is an important and desirable characteristic for virtual environments, particularly virtual learning environments, and propose that the nature of the avatars involved could be a contributory factor in the degree of presence engendered. In order to explore this hypothesis, the series of experiments described below was conducted using a variety of avatar styles.

9.2 Presence Measures

The usability of an interface is defined as a measure of the ease with which a system can be learned or used, its effectiveness and efficiency, and the attitude of its users towards it (Preece *et al.*, 1994). The main difficulty with CVE usability evaluation is the fact that CVEs are founded on very recent technology, and so far only prototypes of truly collaborative three-dimensional virtual learning environments currently exist.

Based on the degree of involvement that has already been observed within full-immersion virtual reality systems, Bricken and Byrne (1993) propose that an obvious benefit of presence in educational CVEs will be that it leads to a greater degree of engagement and excitement on the part of the learners. Considering presence as a result of cognitive and social immersion to be the prima facie "key added value", researchers have, however, only just begun to analyse the nature of presence, what cognitive variables are connected to presence, how presence is generated in multi-user VEs, and what its benefits for education and training might be. Further, as presence characterises the response of participants to the system, it is seen as an obvious choice for the key variable in the usability evaluation process of CVEs. Indeed, presence has been used before as the basis for predicting performance in, and potential benefits of, new learning systems in Sheridan (1992) and Held and Durlach (1992).

Measuring presence is not a trivial task, however. Asking questions that measure only the subject's perception of the technology that contributes to immersion can easily be confused with actually measuring a subject's feeling of "being there", or their behavioural responses to events in the VE. The vast majority of evaluation studies measure presence through questionnaires in an attempt to elicit subjective feelings of presence (Slater and Usoh 1994).

There have been some suggestions for more objective measurement of presence. For example, Sheridan (1992) was concerned with whether subjects duck, blink or carry out other involuntary movements in response to a sudden event. However, there are problems in attempting to infer the effects of the deployment of avatars in Web-based multi-user virtual environments through such a simplistic mechanism. For example, an involuntary response might also be caused by a sudden loud noise, without implying or correlating with a feeling of presence by the user at that time.

The most suitable approach to the measurement of presence is heavily debated among researchers (Slater, 1999; Witmer and Singer, 1999). Within the current experiment the approach to measuring presence largely followed the methodology of Witmer and Singer (1998), who argue that *involvement* and *immersion* are both necessary for experiencing presence. Whereas *involvement* is defined as a psychological state experienced as a consequence of focusing one's attention on a coherent set of stimuli, *immersion* is a psychological state characterised by perceiving oneself to be in an environment of continuous stimuli and experiences.

However, as these presence measures apply only to single-user virtual environments, extended presence measures, namely *awareness* and *communication*, are needed to cover issues specific to multi-user collaborative virtual environments. All four measures (involvement, immersion, awareness and communication) were therefore used in the current research. Furthermore, it has been argued that measuring presence makes sense only when speaking about the degree of presence in one virtual environment setting relative to another (Slater *et al.*, 1998), since presence cannot be measured in absolute quantities. The current research acknowledges this and therefore populates the same world model with different types of user embodiment, thus enabling comparative measurement of presence and hence a meaningful evaluation.

9.3 Hypotheses and Experimental Design

The experimental study aimed to find out whether the appearance of avatars influences the level of presence. To assess this, three types of user representation were constructed: basic shapes, animated cartoon-style avatars, and animated realistic avatars. The basic shape avatars (Figure 9.1) were created in VRML, the animated cartoon-style avatars (Figure 9.2) were created by Avatara (http://www.avatara.com/), and the animated realistic avatars (Figure 9.3) were created by Cybertown (http://www.cybertown.com/).

The implementation of the experiment was fully Web-based. Pre- and post-questionnaires relating to the experience within the virtual art gallery were implemented as CGI/Perl online forms to be submitted by subjects electronically. The virtual gallery model was implemented in VRML and comprised only basic shapes for defining the geometry of the room and the picture frames. The *Blaxxun Community* virtual world server (see http://www.blaxxun.com/) was used to make the virtual gallery accessible on a Web server and enable avatar and chat interaction. The *Blaxxun Contact* VRML browser was used on the client side. Terminals to access the system were provided at locations within Leeds Metropolitan University and Axis premises.

A collaborative task was given to subjects, designed to stimulate interaction and communication. The task involved identifying the art style of a number of contemporary artworks. To simplify the task and to aid subjects without expert knowledge in the visual arts, participants were provided with a list of six different styles to select from – Cubist, Abstract, Naïve, Celtic, Psychedelic and Surreal. Their task was unanimously and collaboratively, as a group, to assign the most appropriate style to each of four artworks with which they were presented (see Figure 9.4 for a typical example). Since the group had to agree on one joint decision, the task was collaborative in nature.

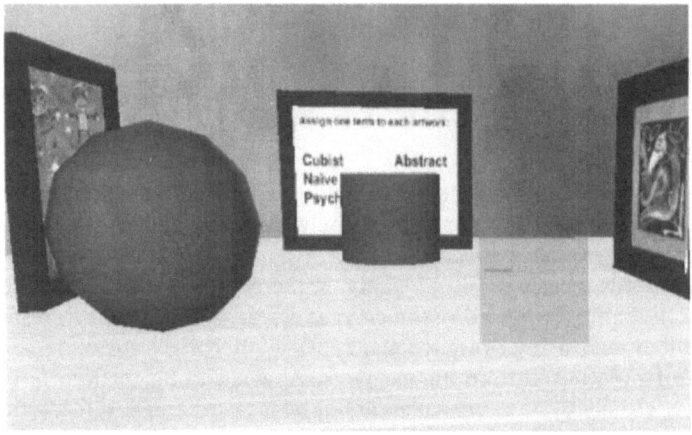

Figure 9.1 Avatar style – Shape.

Figure 9.2 Avatar style – Cartoon.

Twenty-seven subjects took part, their scores on the pre-experiment question-naires being used to divide them into three matched groups of three subjects each that were then randomly assigned the three avatar styles. The various variables involved in the experiment are shown in Figure 9.5. The two variables that need to be controlled are previous *experience* and individual *immersive tendencies*. To reduce the effects of *maturation* (rehearsal effect), another potential source of internal invalidity, a between-group design, was applied to guarantee that subjects participated only once in the experiment.

Figure 9.3 Avatar style – Realistic.

Image	Style	Title	Artist name
	Naive	Silver Haired Children	Mandy Wrightson
	Celtic	Lyre Bird	Christina Scurr
	Surreal	Fantastic Mr Fox	Tomas Lewis
	Psychedelic	Blue Moon Over Marrakesh	Johnny McGuinness

Figure 9.4 Example exhibited artworks from the Axis Database.

The post-questionnaire employed attitude statements with Likert scales as well as open-ended questions to reveal attitudes, beliefs and experiences of subjects (Silverman, 1993). These questions aimed to measure the degree to which aspects

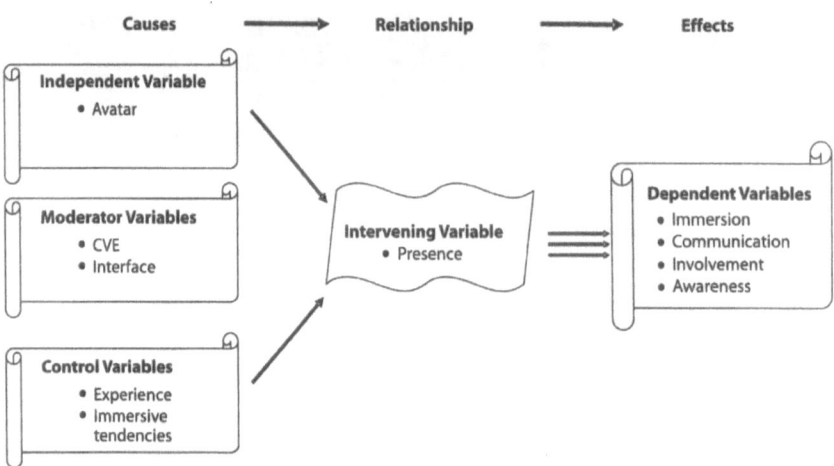

Figure 9.5 Combined variables.

of the virtual environment engendered a sense of presence. The questionnaire collected data regarding the dependent variables of immersion, communication, involvement and awareness, and also covered the moderator variables relating to the nature of the environment itself together with its user interface.

9.4 Results and Analysis

Figures 9.6, 9.7 and 9.8 summarise the data from the experimental study. Results showed that the effects of avatars on presence differed significantly overall ($F_{2,24} = 26.155, p < 0.0005$) between the three groups. Further analysis indicated that the degree of presence was significantly higher when deploying cartoon-style avatars as opposed to basic shape avatars ($p < 0.0005$). Similarly, the degree of presence was found to be significantly higher when deploying realistic avatars as opposed to basic shape avatars ($p < 0.0005$). In contrast, the degree of presence was not significantly different between the use of realistic avatars as opposed to cartoon-style avatars.

These findings were supported by the data collected from subjects by the questionnaire, particularly the open questions. Subjects with cartoon-style and realistic avatars displayed a more positive general attitude towards the experiment and answered in much more detail than those with basic avatars. Furthermore, when directly questioned about avatars their answers were more positive and described the avatars as *amusing, realistic, funny, adding to the experience, interesting* or *excellent*. On the other hand, basic avatars were in some cases not

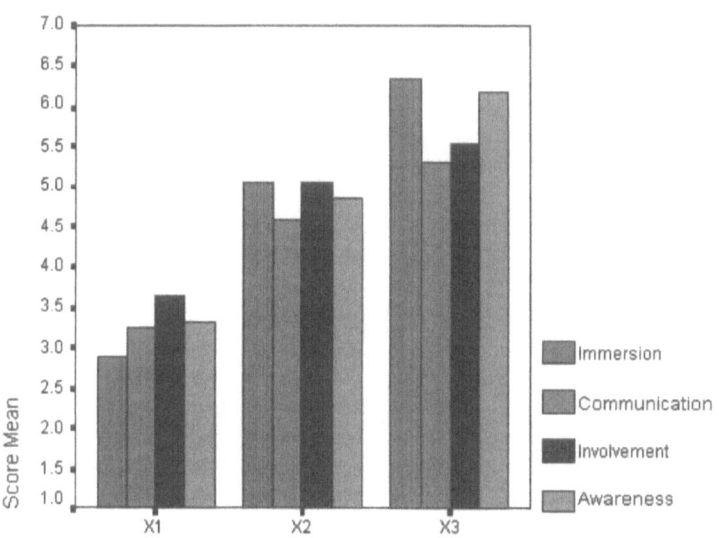

Condition (X1=basic, X2=cartoon, X3=realistic)

Figure 9.6 Subscale scores.

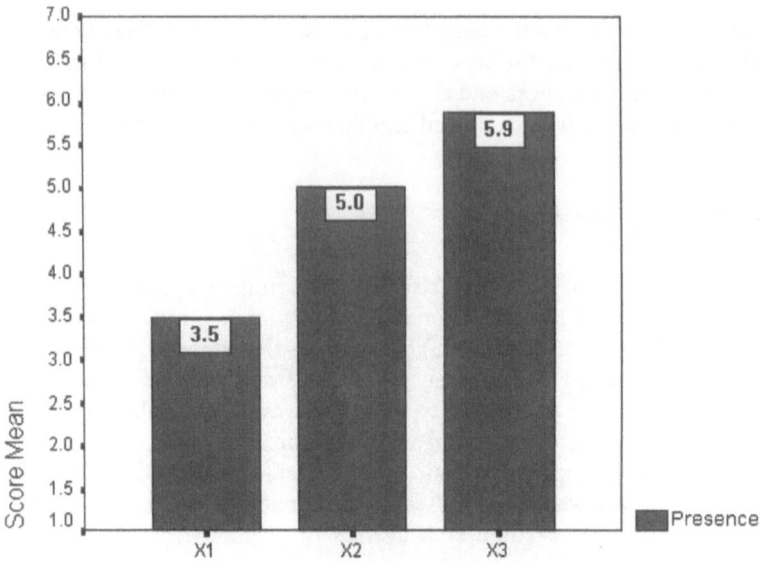

Figure 9.7 Presence scores.

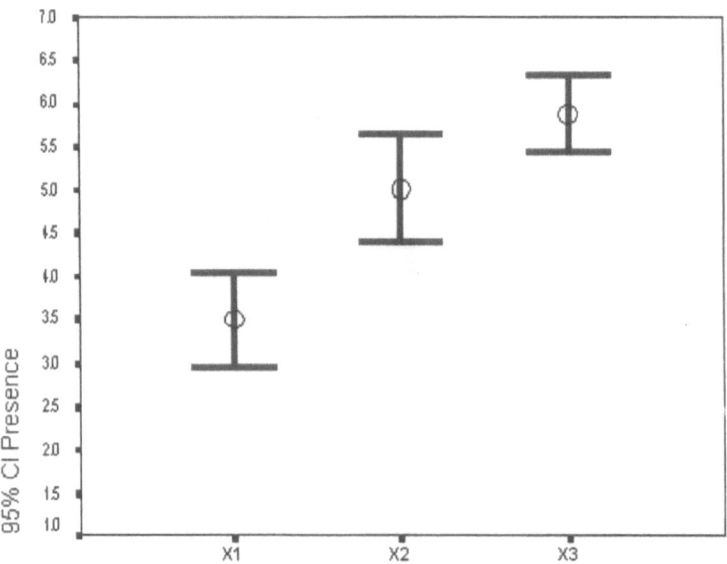

Condition (X1=basic, X2=cartoon, X3=realistic)

Figure 9.8 95% confidence interval for presence means.

recognised as virtual bodies at all and in other cases were referred to as *very poor, could be better, extremely simple* or *could be improved.*

Overall, the results of this study show that different avatar styles do influence presence to different degrees, in particular that animated cartoon-style or realistic avatars promote more presence than basic shape avatars. Further, the former gave rise to sufficiently high scores on the presence scale to strongly suggest that the deployment of such animated avatars does indeed engender a feeling of presence. Additional evidence for this was provided by qualitative data obtained during the experiment and suggested a consequential benefit to the perceived experience within the virtual environment. Thus the experiment succeeded in finding empirical evidence for the benefits of some styles of avatars through measuring the cognitive variable of presence. This evaluation technique will therefore be used again in evaluating the concept of a hybrid avatar/agent model for user representation in educational CVEs, which is the focus of forthcoming experimental work by the authors.

9.5 Discussion

Overall, the statistical results of this study strongly suggest that the deployment of animated avatars improves the CVE experience of subjects with respect to presence; that is, animated cartoon-style or realistic avatars cause more presence than basic shape avatars. It is reassuring that these statistical findings were supported by the qualitative data from the questionnaires.

In sum, this experiment has succeeded in not only evaluating the use of avatars in CVEs, but also in finding empirical evidence for the benefits of animated avatars by measuring a cognitive variable called *presence.* Evidence was found for the hypothesis that the deployment of animated, cartoon-style or realistic avatars improves the virtual experience of participants compared with those represented by basic avatars. Thus, the results of this study strongly recommend employing animated avatars in the design of future educational CVEs.

Further, it can be argued that in a situation of time-independent collaborative learning a continuous presence of all participants is desirable. When members of a group are not co-present, there is a lack of community feeling (Huxor, 1998). A continuous presence of all participants may be achieved using agent technology to control the avatar when the underlying user is not present (Gerhard and Moore, 1998). Thus the success of the current experimental design and evaluation approach is seen as a stepping stone towards evaluating the notion of a continuous presence achievable by a hybrid avatar/agent model, the goal of a forthcoming experiment involving intelligent agents for delivering "presence-in-absence".

References

Bricken, M. and Byrne, C. (1993) Summer students in VR: a pilot study on educational applications in VR technology. In *VR Application and Explorations* (ed. A. Wexelblatt). Academic Press, Toronto.

Gerhard, M. and Moore, D. (1998) User embodiments in educational CVEs: towards continuous presence. In *Proceedings of the International Conference on Network Entities (NETIES '98)*, Leeds.

Gerhard, M., Hobbs, D., Moore, D. and Fabri, M. (1999) Cognitive immersion in CVEs: a hybrid avatar/agent model for user representation in virtual learning environments. In *Proceeding of the Eurographics UK Conference*, Cambridge, UK.

Held, R. and Durlach, N. (1992) Telepresence. In *Presence: Tele-Operators and Virtual Environments*, Vol. 1. MIT Press, Boston, MA.

Huxor, A. (1998), The role of 3D shared worlds in support of chance encounters in CSCW. In *Proceedings of International Conference on Digital Convergence: The Future of the Internet and the WWW*, Bradford, UK.

Kaur, K., Tromp, J., Hand, C., Istance, H. and Steed, A. (1998) Usability evaluation for virtual environments. In *Proceedings of the UEVE '98 Workshop*, Leicester, UK.

Preece, J. (1994) *Human Computer Interaction*. Reading, Addison-Wesley.

Psotka, J. (1995) Immersive tutoring systems: virtual reality and education and training. *Instructional Science*, 23(5–6), 405–431.

Sheridan, T. B. (1992) Musings on telepresence and virtual presence. In *Presence: Tele-Operators and Virtual Environments*, Vol. 1, MIT Press, Boston, MA.

Silverman, D. (1993) *Interpreting Qualitative Data – Methods for Analysing Talk, Text and Interaction*. Sage, London.

Slater, M. (1999) Measuring presence: a response to the Witmer and Singer questionnaire. *Presence: Tele-Operators and Virtual Environments*, 8(5), 560–565, MIT Press, Boston, MA.

Slater, M. and Usoh, M. (1993) Presence in immersive virtual environments. In *Proceedings of the IEEE Conference – Virtual Reality Annual Symposium*, Seattle, USA.

Slater, M. and Usoh, M. (1994) Body centred interaction in immersive virtual environments. In *Artificial Life and Virtual Reality* (eds. D. Thalmann and N. M. Thalmann). John Wiley & Sons, New York.

Slater, M., Steed, A., McCarthy, J. and Maringelli, F. (1998) The influence of body movement on presence in virtual environments. *Human Factors*, 40(3).

Tromp, J. and Benford, S. (1996) Presence, telepresence and immersion: interaction and embodiment in collaborative virtual environments. In *Proceedings of FIVE '95 Framework for Immersive Virtual Environment*, London, UK.

Tuckman, B. W. (1972) *Conducting Educational Research*, Harcourt Brace Jovanovich, Inc., New York.

Witmer, B. and Singer, M. (1998) Measuring presence: a presence questionnaire. *Presence: Tele-Operators and Virtual Environments*, 7(3), 225–240.

Witmer, B. and Singer, M. (1999) On selecting the right yardstick. *Presence: Tele-Operators and Virtual Environments*, 8(5), 566–573.

About the Authors

Michael Gerhard works for AXIS, the national digital repository of contemporary art. He is also a part-time lecturer at Leeds Metropolitan University, and for the last four years his Ph.D. studies have involved him in researching various aspects of online virtual environments.

Dr David Moore, Senior Lecturer at Leeds Metropolitan University, was awarded his Ph.D. in 1993 for work investigating possible designs for an intelligent computer-based dialogue interaction system. He continues to pursue this research interest in computational dialectics, and also has interests in virtual environments, particularly in the potential they may offer in supporting sufferers of conditions such as autism.

Dr Dave Hobbs, Senior Lecturer at the University of Bradford, has for a number of years pursued research in a variety of types of computer-based learning environments. In 1996, whilst at Leeds Metropolitan University, he founded the Virtual Learning

Environments (VLE) research group, which increased the number of postgraduate researchers and publications in the VLE area, and he also co-organised an international conference featuring VLE work.

10

Formative Research on the Refinement of Web-based Instructional Design and Development Guidance Systems for Teaching Music Fundamentals at the Pre-college Level

Wen Hao Chuang

Abstract

Recent advancements in computer and Internet technologies enable universities to implement cost-effective Web-Based Instruction (WBI) and to provide open learning environments 24 hours a day, 7 days a week. This research first synthesised several general WBI design and development guidance systems. Next, the formative research methodology was used to improve that synthesis. A case was chosen for this study, the context being the teaching of music fundamentals at the pre-college level. Efforts were made to identify which guidelines were or were not useful in this case and which guidelines might be beneficial to modify, delete or add in this context. Both on-site and online interviews, observations and document analyses were conducted with all developers involved in this WBI project. As a result of this study, the synthesised general WBI design and development guidance system was revised for best fit with this case.

10.1 Introduction

While more and more WBI courses are continually being developed, little attention is being paid to effective, systemic and systematic WBI design and development.

Table 10.1 Core curriculum at a typical music school.

	1st semester [Fundamentals]	2nd semester
Year 1	Theory and Literature I (T151), Musical Skills I	Theory and Literature II Musical Skills II
Year 2	Theory and Literature III Musical Skills III	Theory and Literature IV Musical Skills IV
Year 3	Theory and Literature V Musical Skills V	
Year 4	History I	History II

Collis (1996) mentions "WWW-based course environments are rapidly appearing, before there has been time for much theoretical development with respect to guidelines for their design" (p. 26). Although there are some general guidance systems focused on WBI design and development process, much of the knowledge about these guidance systems is tentative and lacking in details.

The single case in this study is a project called "Music Fundamentals Online (MFO)". The typical undergraduate core in music schools in the USA consists of a four- or five-semester sequence of parallel courses in music theory and in musical skills (dictation, sight singing, keyboard etc.; see Table 10.1), followed by one to two years of music history. An important part of the core courses is prior mastery of music fundamentals. Although it is important to have basic music skills before entering music schools, many students entering college who plan to have music as their major lack mastery of these basic skills.

10.2 Guidance Systems of WBI Design

WBI is still in its infancy, and there is little research about WBI design and development guidance systems. A number of general-purpose guidance systems for WBI design and development, though general in nature and not usually intended for complex WBI courses, can illustrate both the trends and process involved.

10.2.1 Berge's Guiding Principles in WBI Design

Berge (1998) listed eight guiding principles for WBI design and categorised them into three groups:

- **Pedagogical**
  - Define/describe and list the purpose(s) for each activity, level and type of social and instructional interactivity, and feedback that is desired.
  - Define the levels of teacher control, guided teacher control, student control and group control that are desired regarding each activity.
  - Density of content should be inversely related to the amount of synchronous communication within the Web-based educational learning environment.

- **Technical/support**
  - Recognise that while online environments such as the Web permit multiple media, currently text and graphics are the easiest to use.
  - Use the principle of technological minimalism.
  - Adequate technical support and training for both student and instructor is essential.
- **Social**
  - An important goal of Web-based learning is the creation of an environment of cooperation and trust among students and the instructor.
  - In general, synchronous communication is more expensive than asynchronous. Still, both synchronous and asynchronous modes of communication are important Web-based tools in teaching and learning (Berge, 1998, p. 33).

Among all of Berge's guiding principles, some of them seem to be rather descriptive and lack detailed instruction or action to tell practitioners what to do and how to do it. It will be beneficial to improve these guiding principles in a specific context based on formative research.

10.2.2 Welsh's Event-oriented Design (EOD) Model for WBI

Welsh (1997) first mentioned that any instructional design model for WBI must meet the following criteria:

1. It must be systematic, and therefore useful as a standard online course development methodology.
2. It must be adaptable to different educational disciplines and to differing pedagogical orientations.
3. It must be technology independent, incorporating technologies in wide use for instruction, as well as new technologies such as the Web.
4. It must be useful in traditional contexts, so faculty can recognise the benefits of the design approach in instructional contexts other than WBI (Welsh, 1997, p. 160).

According to Welsh (1997), the EOD model involves consideration of three elements that draw from the fields of distance education and instructional design. These three elements are "asynchronous vs. synchronous learning, specification of performance objectives and the determination of instructional strategies for meeting objectives, and specification of information technologies best suited to meet instructional goals in distance contexts" (p. 160).

In the EOD model, a course is first conceptualised as a series of individual modules. Each module is comprised of a series of instructional events, each of which results in students meeting specific performance objectives. In summary, designing for WBI using the EOD model involves the following steps:

1. Specify instructional goals and performance objectives of the course using traditional instructional design methods.

2. Sequence performance objectives and chunk them into a series of instructional modules, each of which results in students meeting objectives. While instructional modules need not be equal in duration or scope, parallel structuring can establish a comfortable rhythm for the students and instructor.

3. Divide each module into a series of instructional events.

4. For each event, specify event types: full synchronous, limited synchronous, or asynchronous.

5. For each event, specify appropriate Web-based technology to enable the event. Care should be taken to choose only from Web-based technologies available to the instructor and all students.

6. For each event, develop Web-based content where needed and define procedures that ensure smooth completion of the event.

7. Engage in formative evaluation and pilot testing as necessary to verify that each event, as well as the course as a whole, is robust pedagogically and procedurally (Welsh, 1997, p. 162–163).

10.2.3 Gibson and Herrera's Study

Gibson and Herrera (1999) described how a traditional undergraduate classroom-based course was redesigned to online delivery and the several stages of design and development. This study provided the following recommendations (guidance) for the WBI design and development process:

1. Decide upfront if your goal is to simply put some courses online or to design an entire online program. If the former, the resources needed are much less. If you are not sure whether your faculty or administration or even your technical system will support an online program, start by developing a few courses and offering them to current students

2. Use an existing course of studies, hopefully one that you have had much success with so that you are not doing curriculum development and learning how to teach online at the same time.

3. Identify enthusiastic faculty champions right away. Faculty support is the most important element; you cannot succeed without it. We recommend that you choose only full-time faculty at the outset; bringing in outsiders will forever diminish the status of the program to "continuing education".

4. Allocate the financial resources to pay your faculty developers. Online development is very time-consuming, and although you are providing new, marketable skills to the faculty participants, there is an opportunity cost to them.

5. Treat your developers as a team; hold frequent meetings. They need to share ideas and help each other stay focused. There is much frustration during the learning curve. Reinforce their work and recognise their accomplishments at every opportunity.

6. Make sure that technical support is readily available to faculty and students. This includes having the right software and hardware provided to faculty and, most importantly, having technical people ready to help the faculty whenever they may need it. Build in this same level of technical support for students when the classes begin.

7. Do whatever you can to assure that your university has an adequate online library. Students taking online classes are doing so for the freedom from logistical boundaries. The online library services, for example, provide students immediate access to a wide variety of full-text journals as well as e-mail, fax and regular mail access to everything else. (Gibson and Herrera, 1999).

10.3 Methodology

10.3.1 Phase I: Synthesis of WBI Design and Development Guidance Systems

There were two phases in this study: an initial synthesis that focuses on theory creation, and formative research, which focus on theory improvement. During the phase of initial synthesis, all of the guidelines from different WBI design and development guidance systems were kept in their original form but were organised into four categories: technology, pedagogy, implementation and others. Similar guidelines were also grouped together for easier comparison. The initial organisation was suggested by one of the developers to facilitate the brainstorming process during the formal interview sessions. This organisation later was further revised (as suggested by Reigeluth) based on chronological order, as we thought this would help the practitioners more.

10.3.2 Phase II: Naturalistic Formative Research Study and Rationale

The purpose of this phase was to try to suggest improvements for the synthesised guidance system based on the empirical evidence gathered in a single case study, the context being in the teaching of music fundamentals at the pre-college level. The formative research methodology was chosen. Formative research is a type of developmental research or action research that is intended to improve a design theory for instructional practices or processes (English and Reigeluth, 1996; Reigeluth & Frick, 1999). It was originally derived from formative evaluation, which has the primary goal of improving an instructional product while it is being developed in order to achieve the objectives for which it was designed (Dick & Carey, 1996). Reigeluth and Frick (1999) further point out that "for an applied field like education, design theory is more useful and more easily applied than its descriptive counterpart, learning theory" (p. 633). The focus of formative research, therefore, is to improve a design theory (instructional theory) and to provide detailed prescriptive guidance.

10.4 Results

After comparing and triangulating the developers' comments with other data sources plus my own synthesis and observations, a revised set of guidelines was generated. This revised set of guidelines was then sent to each developer for further comments and elaboration. Some of these guidelines were shifted to different phases based on developers' further comments, but no new guidelines were added. The following guidelines are the final results. Note that "New" stands for new guideline generated from the formative research, "Mod" stands for modified guidelines (from original source), and "Orig" stand for original, unchanged guidelines.

Analysis and Planning Phase

- Evaluate all possible instructional solutions. (New)
  - Conduct a survey to see if other schools have similar instructional problems and see if they already found a good solution to the problem.
  - If other schools have found a good solution, evaluate their solution to see if it fits into your own situation. If no schools have a good solution, evaluate other possible solutions.
  - Justify the technologies or solutions you choose.
  - Analyse cost/benefit issues beforehand.
- Assess the readiness of the community. (New)
  - Make sure that the intended audience has the proper equipment and Internet connections to access the WBI course.
  - Make sure that the intended audience will accept WBI as an instructional approach.
  - Make sure that the host institution has the proper network infrastructure to support the WBI course.
- Secure in advance the financial resources to pay your developers. (Mod)
  - Develop a detailed budget to accurately anticipate costs.
- Get support from faculty members and other stakeholders. (Mod)
- Identify enthusiastic faculty champions right away, and get them involved with the project. (Mod)
  - Choose only full-time faculty at the outset.
- Decide upfront whether to implement an entire course online or just selected lessons. (Mod)
- Use an existing course of studies for your curriculum development if possible. (Mod)
- Start by developing small modules and test them at early stages of development. (Mod)
- Conduct a task analysis, and list all required tasks in as much detail as possible. (New)
- Treat your developers as a team; hold frequent meetings. They need to share ideas and help each other stay focused. There is much frustration during the learning curve. Reinforce their work and recognise their accomplishments at every opportunity. (Orig)
- Make sure that the project director knows about current computer/Web-based technology. If not, find an interface person who can explain things to the director and act as a bridge. (New)

- The project director should be one of the developers or part of the development team if possible. (New)
- Make sure that all developers are familiar with the content and have experience teaching with the content area if possible. (New)

Design Phase

- Write down your instructional objectives in detail and list resources required. (Mod)
- When possible, use an existing course as a model to develop your instructional modules to speed up the design process. (Mod)
- Conceptualise a course as a series of individual modules, with each module comprising a series of instructional events. (Orig)
- Sequence performance objectives and chunk them into a series of instructional modules, each of which results in students meeting objectives. When possible, employ *parallel structuring* to help establish a comfortable rhythm for the students. (Orig)
- For each event, specify appropriate technology to enable the event. Care should be taken to choose technologies available to all students. (Mod)
- Create a safe, non-threatening and reliable online learning environment for the learners.
  - Make sure the learners feel comfortable performing at an early stage of learning.
  - Use early encouragement and reassurance to help the learner feel comfortable about making initial mistakes.
  - Build up trust between learners and the online learning system. The system should be reliable enough so that students can trust the online learning technology and don't have to worry about losing their completed tasks.
- Engage in formative evaluation and pilot testing as necessary to verify that each event as well as the course as a whole is robust pedagogically and procedurally. (Orig)
- Define/describe and list the purpose(s) for each activity level and the type of social and instructional interactivity and feedback that is desired. (Orig)
- Define the levels of teacher/computer control, student control and group control that are desired regarding each activity. (Mod)

Development Phase

- Build up a knowledge-sharing and proactive working culture and promote innovation in your development team. (New)
- Evaluate and choose adequate course authoring/development tools at early development stages. (New)

- Hire courseware developers who are familiar with the course authoring/development tools you choose. (New)
- Build a simple group Intranet to share design documents among team members. (New)
- Use an Instant Messenger program (or other communication tool) when necessary for better team communication. (New)
- Subscribe to or monitor newsgroups/listservs of the development tools you are using. (New)
- Make sure that staff engagement and commitment are happening in your team, as it is very important to the success of the WBI design and development process. (New)
- Use the minimum technology required to achieve the instructional objectives. (Mod)
- Keep the online media (e.g. multimedia files) size as small as possible. (Mod)
- Develop a subsystem in your WBI to capture each student's problem solving process. This subsystem should be able to: (New)
  - Keep a history of students' correct answers.
  - Determine if mastery has been reached. The mastery criteria can be made of three parts:
    - a minimum number of problems that must be attempted
    - a maximum history list length (often the same as minimum number of problems to try)
    - a minimum percentage correct of the problems being counted
- Test your prototype on multiple platforms and browsers at early stages of development. (New)
  - Be aware that Web pages on different browsers and different platforms can look very different.

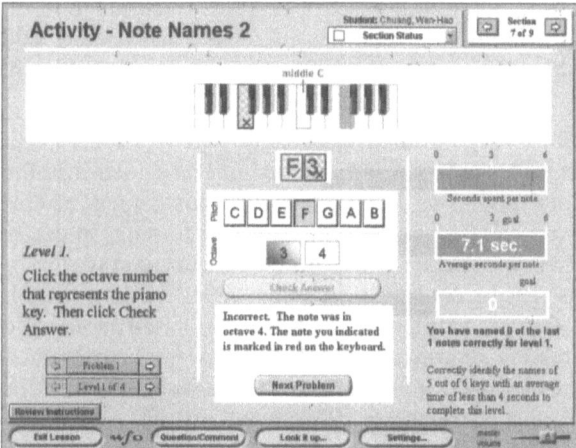

Figure 10.1 The prototype of MFO.

  – Identify any cross-platform compatibility problems as early as possible.
- Provide adequate technical support for students in a Web-based instruction environment. (New)
  – Provide email or telephone technical support to answer students' technical questions.
  – Build up a knowledge base for the most frequently asked technical questions.

Figure 10.1 shows the prototype of "Music Fundamental Online (MFO)".

10.5 Conclusions

It was not possible to examine all aspects of WBI design and development guidelines and generate a perfect guidance system by synthesizing several general WBI design and development guidance systems and analysing empirical evidence in a single case study. Therefore, this study is just a beginning, and more studies are needed to further confirm and elaborate the findings in this study. Schools that are interested in WBI should analyse every possible solution before making the decision, as WBI might not be the best solution in a lot of situations and learning domains. Most of the media choices are done not for instructional purposes, but for implementation purposes (Powell, 2000). Also, every WBI needs to be supported by instructional theory. According to Kulp (1999), the most common genres in WBI courses are learner-centred topics. Students interact with material in the course Web site in somewhat the same way they would with a computer-based training (CBT) self-study course (Kulp, 1999). They actively interpret information and experience in order to create new knowledge or build new work products of some type. Based on this assumption and Gordon's (1997) "Music Learning Theory", we can hypothesise that most of the guidelines supported by this case can also be useful in creating other skill-based WBI courses, although more research is needed to confirm this hypothesised generalisation. For other skill-based pre-college remedial WBI courses, the guidelines provided in this study might also be useful.

Secondly, in this study, not many guidelines were rejected. This indicates that most guidelines were useful based on the empirical evidence for this case. Some guidelines were refined to better fit in with this context of creating a WBI course at the pre-college level. One of the major findings in this study is additional guidelines and further elaboration of the original synthesized guidance system in this specific context. Also, some "holes" in the original synthesised guidance system were found, and new guidelines based on the experience from this case study were identified to fill those holes. But note that in a single case, there can be no evidence for further generalisation; therefore the results cannot be generalise beyond this case. Only hypothesised generalisations can be offered, as mentioned above.

References

Berge, Z. (1998) Guiding principles in Web-based instructional design. *Educational Media International*, 35(2), 72–76.

Collis, B. (1996) *Tele-Learning in a Digital World: the Future of Distance Learning*. International Thomson, London.

Dick, W. and Carey, L. (1996). *The Systematic Design of Instruction*. HarperCollins, New York.

English, R. E. and Reigeluth, C. M. (1996) Formative research on sequencing instruction with the elaboration theory. *Educational Technology Research & Development*, 44(1), 23–42.

Gibson, J. W. and Herrera, J. M. (1999) How to go from classroom based to online delivery in eighteen months or less: a case study in online program development. Available at http://www.thejournal.com/magazine/vault/A2080.cfm.

Gordon, E. E. (1997) *Learning Sequence in Music: Skill, Content, and Patterns: A Music Learning Theory*, 1997 edition. G.I.A. Publications, Chicago, IL.

Kulp, R. (1999) Effective collaboration in corporate distributed learning: ten best practices for curriculum owners, developers and instructors. *IBM Learning Services*, November.

Powell, G. C. (2000) *Are You Ready for WBT?* Available at: http://itech1.coe.uga.edu/itforum/paper39/paper39.html.

Reigeluth, C. M. and Frick, T. W. (1997) Formative research: a methodology for improving educational theories and models. In *Instructional-design Theories and Models: a New Paradigm of Instructional Theory* (ed. C. M. Reigeluth). Lawrence Erlbaum Associates, Mahwah, NJ, pp. 645–650.

Welch, R. and Frick, T. (1993) Computerised adaptive mastery tests in instructional settings. *Education Technology, Research & Development*, 41(3), 47–62.

Welsh, T. (1997) An event-oriented design model for Web-based instruction. In *Web-Based Instruction* (ed. B. Khan). Educational Technology Publications, Englewood Cliffs, NJ.

11

Unnatural History? Deconstructing the *Walking with Dinosaurs* Phenomenon

Anne M. White and Karen D. Scott

Every age creates prehistory in its own image

Adams (1999)

11.1 Introduction

In their book, *Remediation: Understanding New Media* (1999), Jay David Bolter and Richard Grusin argue that digital visual media can best be understood by examining the ways in which they relate to earlier technologies of representation, a relationship which is a complex mix of homage, critique and revision. They also suggest that this process, which they refer to as "remediation" takes two main forms: immediacy, in which the presence of the medium is downplayed in order to achieve an effect of transparency and realism, and hypermediacy, in which the medium is foregrounded and draws attention to its own artificiality.

In this chapter, we explore Bolter and Grusin's ideas using the BBC series *Walking with Dinosaurs*, the related documentary entitled *The Making of "Walking with Dinosaurs"* and the official supporting BBC Online Web site. Firstly, we focus on the process of remediation, by analysing the various ways in which *Walking with Dinosaurs*, one of the first television series to fully exploit the potential of digital technology, utilised codes and conventions from earlier media forms and technologies. We will examine whether the overall effect achieved in these programmes can best be described as immediacy or hypermediacy, and having examined the television series in depth, we will contrast this with the associated documentary and Web site. We will conclude by exploring some of the broader issues relating to media representation which are raised by this ground-breaking television concept.

11.2 The Context

Like Bolter and Grusin, we believe that "No medium today, and certainly no single media event, seems to do its cultural work in isolation from other media, any more than it works in isolation from other social and economic forces" (Bolter and Grusin, 1999, p. 15). We will begin, then, by contextualising the television series that will form the major focus of this chapter.

It would be true to say that in the year marking the countdown to the start of the new millennium, the BBC did not appear to be in the best of health. During 1999, the Corporation had seen some of its star talent and top executives defect to commercial television (Robins, 1999) and it was involved in a major battle with ITV about the latter's poaching of some of its most highly successful programme formats (McCann, 1999). There was a continuing debate about whether standards of public broadcasting were slipping (Jury, 1999) and widespread discussion about the wisdom of the Corporation's decision to expand into the new media. All of this caused one journalist to suggest that the acronym BBC might be more accurately rendered as "Barren, Banal and Confused" (Robins, 1999). In mid-September 1999, the viewing figures for BBC1 were registered as being at their lowest for two years (Gibson, 1999a), but less than a month later, a six-part documentary series began which was to breathe new life into the schedules of the ailing channel: its title, *Walking with Dinosaurs.*[1]

The cover of the Press pack confidently proclaimed that it was to be "The biggest thing on television in 200 million years" – certainly it was the most expensive documentary series the BBC had ever been involved with, at a reported cost of some £6 million pounds.[2] Everything about the series was to be the biggest and best: it would show "the most spectacular creatures", and would make use of "the latest scientific thinking" and "state-of-the-art" technology including advanced computer animation and animatronics (Press pack). An impressive array of academics and technical experts had been involved in the project which took over two years to bring to fruition. Seven palaeontologists had been employed as scientific advisors and more than 100 other academics from various fields had been consulted. Over a dozen animators from the award-winning company FrameStore had worked together with a small army of sculptors, animatronics experts, sound designers and location crews to transform the original idea into televisual reality. The various contributions which all these groups had made to the creation of the series was explained in a 50-

---

1   At least two journalists drew parallels between the subject matter of this series and aspects of the dire situation in which the BBC found itself: Lawson (1999) commented that "The fact that the BBC is seeking to recover from a period of bad publicity and uncertainty over its role with a programme about an all-powerful force which became extinct is unlikely to have been missed by senior executives", whilst Adams (1999) noted "*Walking With Dinosaurs* may have ostensibly been about the ancient past, but its real purpose was to suggest that in the digitalised (*sic*) television future the Beeb was not prepared to be easy prey for circling rivals scenting blood".

2   *Walking with Dinosaurs* was a BBC/Discovery/TV Asahi co-production in association with ProSieben and France 3.

minute documentary, *The Making of "Walking with Dinosaurs"*, which was screened along with the series in October 1999.

According to the producer of the series, Tim Haines,[3] the programmes represented "the world's first Natural History of Dinosaurs" and the truly innovative feature of the series was that it would provide viewers with "a window into a lost world", allowing them "to believe that they [were] watching living, breathing creatures in their natural habitat" (Press pack). Another innovation was to have a BBC Online Web site, `http://www.bbc.co.uk/dinosaurs/`, linked to the series.[4] Accessible from mid-September 1999 (and still available in February 2002), information was added to the Web site week by week as the programmes progressed, to create a vast educational resource. As well as articles written by experts, and facts and figures about the dinosaurs featured in the series, the Web site had more interactive aspects, such as games and a forum where visitors could ask questions or make comments. In addition, there were three chances to chat live online to experts and those involved in the making of the programme. A BBC book to accompany the television programmes was the first of what was to become a long list of merchandising spin-offs.[5]

Peter Salmon, the controller of BBC1, had predicted that the series would "dazzle audiences with the breadth of its imagination and the quality of its scholarship" (Press pack). Certainly, if viewing figures are to be believed, then *Walking with Dinosaurs* was an astonishing popular success. According to BARB (Broadcasters' Audience Research Board), there was an audience of 15 million for the first episode broadcast on 4 October 1999, with another 3.91 million tuning in for the repeat on the following Sunday, making it by far the most watched science programme in British television history (*The Guardian*, 1999c). The reception from television critics, however, could best be described as mixed. At one extreme there was wild enthusiasm (Boucher, 1999; *The Guardian*, 1999a; Matthews, 1999; Steel, 1999; Viner, 1999) at the other, utter contempt, with the reviewer Jacques Peretti dismissing the programme as "A high-tech Sooty show" (Peretti, 1999, p. 2).

Many others, though clearly impressed by the glossy production values, expressed minor or major reservations about the way in which the series presented mere speculation as scientific certainty (Banks-Smith, 1999a; Brown, 1999; Hanks, 1999; Kellaway, 1999; Lawson, 1999; McKie, 1999). Perhaps the most damning criticism was to be found in a newspaper article written by Dr Paul Barrett, a palaeontologist from Oxford University who had been consulted on his specialism for the series. In his opinion, the BBC had "missed an excellent opportunity to produce a world-class science programme, and gone for the softer option of making a dinosaur soap opera instead" (Barrett, 1999, p. 8).

---

3   Haines joined BBC Science in 1988 and his credits included a stint as producer for the popular science programme, *Tomorrow's World*. He also has a degree in zoology.
4   Marketing information from the BBC claims that the UK site had nearly a million visitors in the first week of transmission of the series.
5   According to Gibson (1999b), the marketing and licensing deals related to the series were reportedly expected to make as much as 60 million pounds for the corporation. See also Horrie (1999).

There was also a certain amount of cynical comment from media journalists like Banks-Smith (1999b), who suspected that the possibility to exploit the appeal of the series for younger viewers would not be lost on the BBC World-wide executives responsible for the licensing of merchandising tie-ins: "I begin to think", wrote the journalist that "the whole thing is geared to selling chocolate dinosaur eggs to five-year-olds".[6]

11.3 Key Concepts of Remediation

Having situated *Walking with Dinosaurs* within its general cultural and socio-economic context, we will now move on to consider some of the key aspects of Bolter and Grusin's ideas about how we make sense of new media, in particular the concept they refer to as remediation. They argue that in order to understand more fully the cultural significance of new media technologies, there is a need to focus on the nature of their relationship with earlier technologies of repre-sentation and already existing media forms. Moreover, they believe that by examining the various ways in which these new and old media interact with each other, valuable insights can also be gained into how we relate to these different kinds of media and interpret the differing ways in which they mediate reality.

For Bolter and Grusin, this relationship between media technologies is to be understood as a two-way, interactive process of recombination, reconfiguring and refashioning in which technologies borrow freely from each other and media forms are endlessly recycled. They refer to this constant interplay between new and existing media technologies as remediation and although they argue that it can be considered to be one of the defining features of the new digital media which emerged in the latter part of the twentieth century, they also point to its widespread use as a cultural practice in Western countries, and to its long history, which they illustrate by means of a wide variety of examples. Moreover, they examine how the process of remediation is directly related to the history of representation itself and to the twin styles or strategies of repre-sentation which have emerged over the course of the centuries, namely: (trans-parent) immediacy and hypermediacy. Since these concepts are of direct relevance to our case study, we will say a little more about them here. For the sake of brevity, we have included an overview of some of the key characteristics which are identified as being typically associated with each of these styles (Table 11.1).

Is often referred to as duping viewers into believing they are looking through a window onto the world Normally thought of as alerting viewers to how the illu-sion of reality is created

---

6   Despite the very disparate subject matter, certain parallels can be drawn between these concerns and those expressed about the phenomenal success of another innovative BBC series, the children's television programme, *Teletubbies*; see White (1999).

Table 11.1 Immediacy and hypermediacy (based on Bolter and Grusin (1999)).

Immediacy	Hypermediacy
Ignores/denies the presence of the medium	Draws attention to the presence of the medium
Conceals mediation process and producer	Foregrounds mediation process and producer
Transparency	Opacity
Seamless	Fragmentary
Unified space	Heterogeneous space
One point of view	Multiplicity of points of view
Medium made to disappear	Medium foregrounded

Successive developments in visual media, like painting, photography, film and television, have all claimed to offer the means of better satisfying our desire to recreate the illusion of immediacy, of providing us with what is commonly referred to as a window on the world. Indeed, each new medium is normally sold to us as an innovation on the basis that it is an improvement on its predecessors precisely for this reason: it will provide more natural colour, more life-like images, etc. When the new medium becomes a serious rival for the sociocultural prestige formerly associated with the older media, and/or a probable competitor in economic terms, the more traditional media typically respond by attempting to refashion or remake themselves in its likeness, by imitating and incorporating wherever possible aspects of its innovative features but without drawing attention to the source. Thus, for example, Hollywood films now routinely make use of digital compositing techniques to remove unwanted elements from scenes involving special effects or stunts, erasing anything in the finished product which threatens to disrupt the illusion of immediacy for the film viewer.

However, there is also another style of representation with an equally long tradition which Bolter and Grusin refer to as hypermediacy, a style in which the aim is to draw attention to the medium itself, deliberately highlighting the fact that what we view is not a transparent window on the world but merely a mediated representation. The idea of this form of representation is, in short, "to make the viewer acknowledge the medium as a medium and delight in that acknowledgement" (Bolter and Grusin, 1999, p. 42) and the fragmented heterogeneity of a new media text, such as a World Wide Web page, could be considered to be the epitome of hypermediacy.

Bolter and Grusin argue, however, that closer examination reveals that these two styles of representation often coexist within media forms, and that effectively what we find is that many visual technologies display a tendency to "oscillate between immediacy and hypermediacy, between transparency and opacity" (Bolter and Grusin, 1999, p. 19). Moreover, it is this oscillation, this switching between styles, which these theorists believe can provide "the key to understanding how a medium refashions its predecessors and other contemporary media" (Bolter and Grusin, 1999, p. 19). In our analysis we hope to demonstrate exactly how this oscillation works in practice and what it can tell us about the dynamics of the relationship between television, film, computer animation and the WWW as they are played out within the *Walking with Dinosaurs* texts.

11.4 *Walking with Dinosaurs* – the Television Series

There can be no doubt that the key influence on Haines as producer of the television series was the genre known as Natural History documentary, with the Press pack dubbing the series "the extinct *Life on Earth*". The comparison makes an intentionally humorous reference to an earlier critically acclaimed BBC series (first shown in 1979), but also attempts to position *Walking with Dinosaurs* within this tradition of high-quality wildlife programme-making epitomised by the work of David Attenborough.[7] It is worth mentioning, however, that the BBC discourse surrounding the series also acknowledged the fact that for the vast majority of television viewers the more obvious cultural reference point for *Walking with Dinosaurs* would have been the Hollywood blockbuster movie *Jurassic Park* (1993) and its sequel *The Lost World* (1997).[8] Thus, for example, the article which appeared in *Radio Times* (the BBC's official listings magazine) to accompany the screening of the first episode of the series was entitled "Jurassic Parklife" (2–8 October 1999)[9] and *The Making of "Walking with Dinosaurs"*, the behind-the-scenes documentary which demonstrated how academic expertise had been used to ensure the scientific rigour of the series, contained allusions to Spielberg's work.[10] So although the series was officially labelled as Natural History documentary,[11] it was also linked to a popular cinematic tradition of representing prehistoric life, effectively meaning that it was positioned between two sets of codes and conventions, relating to different genres and different media technologies. As indicated

---

7   There seems to be some evidence that *Walking with Dinosaurs* could be classed as an affectionate homage to earlier BBC wildlife programmes. Thus, for example, the opening sequence of Episode 3: *Cruel Sea*, in which a monstrous marine reptile, liopleurodon, unexpectedly pounces on a dinosaur, bears an uncanny resemblance to one of the most memorable incidents from another of Attenborough's triumphs, *The Trials of Life* (first shown in 1990) in which a killer whale suddenly surges out of the waves to gulp down an unsuspecting seal. The distinctive movements and sociable behaviour of the meerkats made famous in the documentary *Meerkats United* (originally broadcast on 20 December 1993) are also clearly the inspiration for the Leaellynasaura colony in Episode 5.

8   As if to emphasise this fact, *Jurassic Park* was broadcast on BBC1 only two days before the first episode of *Walking with Dinosaurs* was screened.

9   This kind of discourse was also reflected in other newspaper articles and reviews relating to *Walking with Dinosaurs*, in which journalists drew constant comparisons – both implicit and explicit – between the factual television series and Spielberg's filmic fantasies.

10   Although the director's creations are never referred to explicitly, statements in *The Making of "Walking with Dinosaurs"* such as "You won't see *our* diplodocus eating from the treetops" are clearly intended to encourage the audience to draw comparisons between the factual (BBC documentary) and the fictional (Hollywood film).

11   For more on documentary as a television genre, see Kilborn and Izod (1997).

above, this mixing of fictional and factual genres appears to have made many critics rather unsure, at least, about how to approach the programmes. *Walking with Dinosaurs* drew upon already existing forms of televisual and filmic representation and refashioned them via digital technology to produce a new kind of hybrid genre, a remediation which both celebrates and critiques the ways in which earlier technologies of representation have attempted to portray reality.

With Natural History documentary as a genre, the key word with regard to subject matter is, of course, "Natural", for it is typically expected that programmes of this genre will focus on living things (most commonly animals or plants) in their wild state. Thus when, for the first time, a series of this kind takes the extraordinary step of focusing on an extinct life form, it raises a whole series of fascinating issues about what the meaning of immediacy and hypermediacy in this context might be. Bolter and Grusin's observation about how viewers judged the "reality" of the computer-animated creatures in *Jurassic Park* and *The Lost World* is relevant: "Because no one has ever seen a living dinosaur, the viewer is invited to measure the graphics by what she regards as plausible for such huge animals, although her sense of plausibility comes from other films and fiction" (Bolter and Grusin, 1999, p. 154). Haines, in making *Walking with Dinosaurs*, realised that in order to create the sense of immediacy that he wanted – what he called his "window into a lost world" – he needed to ensure that his dinosaur protagonists did not depart too radically from their digital predecessors as envisioned by Spielberg, since this image was still fresh in the minds of viewers.[12]

More importantly, however, Haines understood that viewers would also measure his televisual treatment of an extinct life form against the model of immediacy that had come to be associated with Natural History documentary as a genre and in making the series he deliberately set out to imitate aspects of this, wherever possible.[13] The series used the normal narrative conventions of such programmes, following the patterns established by particular natural cycles, such as those of the changing seasons or of individual life forms, showing their birth, growth, reproduction, maturity and death. The images were also accompanied by a voice-over which used the typical linguistic features of the genre: an authoritative commentary by an omniscient narrator, combining the "objective" discourse of scientific knowledge (facts and figures) with touches of anthropomorphism. The choice of the well-known actor and film director Kenneth Branagh as narrator provided further evidence of the high-profile status which the BBC accorded the series.[14]

Since historically such programmes have been shot on location, showing the living things interacting with their environment, most probably as part of a larger ecosystem, a number of suitable habitats were found and filmed as the backdrop to the various episodes in the series. Specially created animatronic

---

12   Indeed at one time, a technical team from George Lucas's Industrial Light and Magic was being considered to recreate the special effects seen in Spielberg's film, but proved too costly. See *The Guardian* (1999b).

13   For more on the subject of human fascination with images of animals generally, see Berger (1980) and Williams (1999).

models of the creatures to be included in the various episodes were filmed in these locations so that they could be used for close-ups. Meanwhile the computer animation team at FrameStore worked on the digital dinosaurs for the series, modelling these on archive footage of living animals. The same company was also responsible for the final process of compositing, described in the booklet published to accompany the series as "the marrying together of real and virtual footage into a seamless whole" (*Radio Times*, 1999, p. 49) and at first sight, this is exactly the impression that television viewers get. However, on closer examination, each episode proves to be, like the series' hypermediated Web site itself, a particularly intricate collage of fragments of disparate representations, including numerous clips from earlier wildlife documentaries showing real creatures believed to have formed part of the dinosaurs' ecosystem.[15] The complex interaction between these different mediations and the varying degrees of authenticity that they represented proved confusing for at least one experienced television critic, as his description demonstrates: "Computer graphics generate the water and vegetation of the planet at that time while animatronic models reproduce the dinosaurs. [...] Winged creatures fly above what seems to be real water before landing in a tree" (Lawson, 1999).

Perhaps it could be argued that the Natural History programme format was ideal for an experiment of this kind, since as a television genre it has always been quicker than most to make use of new technological developments to provide viewers with access to what would usually remain inaccessible. Indeed certain wildlife documentary series have made a particular feature of the fact that they are offering viewers a privileged, almost voyeuristic, glimpse of worlds that would normally remain hidden, the most well-known recent example being Attenborough's *Private Life of Plants* (1995). Over the course of time, these new visual technologies have impacted on what viewers are willing to accept as being authentic in the context of the representation of the natural world. Thus time-lapse or slow-motion sequences, infrared or heat-sensitive imaging, the extreme close-ups afforded by macro photography – all of these have become accepted means of portraying the reality of the plant and animal kingdoms, even though they show aspects of Nature that would not normally be visible to the naked eye.

Given that, as Bolter and Grusin put it: "Whenever one medium seems to have convinced viewers of its immediacy, other media try to appropriate that conviction" (1999: 9), it was only to be expected that Haines would borrow heavily

---

14 One might also read this use of Branagh as an intriguing intertextual allusion. Much was made in both the popular Press and in the *Making of "Walking with Dinosaurs"* of the idea that Haines and his technicians were bringing the dead back to life, so who better to comment on the successful outcome of this process than the man who had directed *Mary Shelley's Frankenstein* (1994) and starred in this film as the scientist seeking the secret of reanimation?

15 Examples of footage of living creatures include clips of insect life (dragonflies, damsel flies, dung beetles, tree grubs, butterflies); sea creatures (jellyfish, shoals of fish, sharks, horseshoe crabs); plant life (flowers, ferns); reptiles (snake) and the wildlife of the African savannah which concludes the final programme.

from this televisual tradition in representing his own computer-animated creatures in order to convince viewers of their immediacy. This is the aspect which
distinguishes these digital creations from their filmic predecessors and also
proves to be most interesting in the context of remediation. There are
numerous instances in the series when the dinosaurs are shown on screen in a
way which imitates typical wildlife documentary styles. Thus the scenes
involving representations of dramatic life-and-death encounters between
predators and prey, such as a chase between some utahraptors and an
iguanodon (Episode 4), are made to look as though they have been filmed in
slow motion. In Episode 6, comments relating to the physiology of the
tyrannosaurus are accompanied by what appears to be a visual of the dinosaur
produced by heat-sensitive imaging technology. Instances like these not only
imitate particular televisual styles associated with the wildlife documentary,
but might also be said to celebrate certain distinctive aspects of older technologies of vision.

However, perhaps the most interesting aspect of the series is the way in which it
also chooses to reproduce the shortcomings and flaws associated with traditional media in an attempt to convince viewers of the immediacy of what they
are viewing.[16] Thus, for example, in Episodes 1 and 5 respectively, footage of
cynodonts and leaellynasaura which has supposedly been shot under cover of
darkness has the grainy, imperfect monochrome quality of night-time footage.
There are also a number of instances when the viewers' attention is apparently
drawn to a particular shortcoming of the technology required for filming:
namely, that the camera lens is not always able to offer us the flawlessly transparent window on the world that it should. Thus in Episode 2, the image appears
to go cloudy as the hot breath of a meat-eating utahraptor steams up the lens,
whilst in Episode 6, as the camera moves in for an extreme close-up, the roaring
tyrannosaurus seems to shower it with saliva, which again obscures the lens.

Analysis of these two incidents reveals the complexity of the interplay here
between immediacy and hypermediacy. The viewers' initial impression that
this is unmediated reality is proved false when their attention is drawn to the
camera lens and from there to the camera which, it would appear, has been
filming the scene. However, at the same time, the interaction between the
subject being filmed and the camera lens seems to add a different kind of
authenticity to the scene: the camera was apparently so close to the action that
it became part of it. Then, immediately, viewers are forced to rethink these
representations of close encounters with dinosaurs and to recognise them as a
juxtaposition created by digital manipulation. For paradoxically, the moment
that seemingly promises viewers the kind of authenticity they most desire in a
wildlife programme is, in fact, the moment at which they are forced to acknowledge its status as mere fabrication: a perfect example of the oscillation between

---

16   This forms an interesting contrast to the way those working in computer graphics often
      operate, since experts in this area "do not in general imitate 'poor' or 'distorted' photo-
      graphs [...] precisely because these distorted photographs, which make the viewers
      conscious of the photographic process, are themselves not regarded as realistic or
      immediate" (Bolter and Grusin, 1999, p. 28).

immediacy and hypermediacy, between transparency and opacity. It is perhaps not surprising, then, that the scene with the spitting tyrannosaurus should have become the most repeated image of the series, for it seems to symbolise what Bolter and Grusin identify as "the twin preoccupations of contemporary media: the transparent presentation of the real and the enjoyment of the opacity of media themselves" (Bolter and Grusin, 1999, p. 21).

11.5 *The Making of "Walking with Dinosaurs"* – The Documentary

Reflecting on the reasons why viewers are so enthralled by Hollywood films like *Jurassic Park*, which make use of digital special effects, Darley (2000, p. 115) commented:

> It is the bizarre nature of the imagery, rendered so faithfully, that [...] *denies and simultaneously points to* the highly sophisticated artifice involved in its production. It is *both* the bizarre and impossible nature of that which is represented and its thoroughly analogical character (simulation of the photographic) that fascinates, produces in the viewers a "double-take" and makes him or her want to see it again, both to wonder at its portrayal and to wonder about "just how it was done" [emphases in original].

The Making of "Walking with Dinosaurs" (first broadcast in October 1999, shortly after the first episode of the series) can be read, on the one hand, as the BBC's attempt to respond to that viewer curiosity, the desire on the part of the audience to know exactly how the illusion of televisual immediacy was created. The documentary functions as a showcase for the talents of those who were involved in making the series and foregrounds a process of media creation in which technology such as computers and cameras take centre stage. At the same time, however, the programme is also clearly intended to be a means of validating the series' claim to be fact-based documentary realism, and extensive footage of paleontologists and other expert witnesses explaining how various aspects of current scientific research have influenced decisions concerning how prehistoric life was portrayed in the series is used as a means of authentication.

There are many sequences in the programme in which viewers see computer animators and academics engaged in dialogue and these can be read as an attempt to suggest that *Walking with Dinosaurs* represents the successful integration of these two very different discourses. Closer analysis reveals, though, that like the series itself, the documentary oscillates between the twin states of remediation: it makes extensive use of the established codes and conventions of immediacy but also alerts viewers to its status as a hypermediated textual construction. However, whereas the unresolved contradiction which lies at the heart of the series is only occasionally exposed, as though this were a momentary, inadvertent slip, it is openly celebrated in *The Making of "Walking with Dinosaurs"*, a postmodern text which not only plays with established codes and convention relating to form and content but also draws attention to its own status as a remediation in a highly self-conscious fashion from the very outset.

Immediately following the spectacular effects of the opening sequence originally used for the series, an animated utahraptor appears dragging the words
"The Making Of" onto the screen underneath the usual title. This is followed by
the typical images viewers might expect in a documentary about the making of
a wildlife series: a production crew (who are the real production crew of
Walking with Dinosaurs) are seen setting up their equipment and preparing to
film on location. The cameras roll as a tyrannosaurus and her young come into
view, but the programme-maker (later revealed to be Tim Haines himself) stops
the filming. Unhappy about the scene, he offers some directorial guidance to his
prehistoric leading lady before commencing a second take.

As even this very brief sequence illustrates, the documentary (or perhaps this
might more accurately be termed "mockumentary") initially establishes a
directly parodic relationship with the series as the new version of the opening
credits deflates the rather overblown pomposity of the original. However, the
scene which follows suggests a rather more complex intertextual relationship
between series and documentary, since it playfully exposes the bogus claims to
immediacy offered to viewers by *Walking with Dinosaurs* (and, one might
argue, by all wildlife documentaries). Indeed both computers and television
cameras feature prominently throughout the documentary, as if to emphasise
that they, in fact, are the real stars of the show. Older media representations of
prehistoric life forms (clips from *The Lost World* (1926) and an unidentified
television documentary) are used as evidence of how previous attempts at
capturing the reality of these creatures have failed.

For the purposes of our argument here concerning remediation and the ways in
which media technologies interact with each other, one of the most interesting
aspects of this television programme is its use of features which are reminiscent of the hypermediated style more typically associated with the New Media,
in particular the aesthetics of the World Wide Web. As a medium, television has
been described as "the greatest synthesizer, turning to its purposes features
drawn from all previous media" (Leiss *et al.*, 1990, p. 96) and it has always
borrowed heavily from other media rivals, constantly adding new visual styles
to its own repertoire with the intention of persuading viewers of its immediacy.
It is still this quality, identified by Flitterman-Lewis as its "peculiar form of
presentness – its implicit claim to be live" (cited in Bolter and Grusin (1999, p.
188)), that constitutes television's particular claim to superiority over other
traditional media forms such as film, for instance. The rise of "reality television" and the phenomenal success of *Big Brother* (2000) are evidence of the
continuing popular fascination with this aspect of the medium.

However, as viewing figures continue to drop as a result of the proliferation of
forms of home entertainment, television executives are also aware of the need
to retain the interest of the younger audience who are more used to the
windowed and multimediated look of the computer screen. Not surprisingly,
then, there seems to be a deliberate attempt made in *The Making of "Walking
with Dinosaurs"* to replicate the distinctive appearance of this new medium on
several occasions by splitting the screen to show three images at the same time
on screen. However, since in every case these are interrelated, what viewers see

is a much less radical form of montage than would be typical of the usual frag-mented heterogeneity of a computer screen which might combine written text with video clips, still photographs and graphics. At other times, the television screen is filled entirely by the digital animations generated by the computer, and the two separate media spaces appear to have converged completely. On several occasions, however, there seems to be a conscious attempt by the programme producer to disrupt this illusion of convergence by ensuring that there is a visible reflection of the computer animator in the computer screen, so that viewers see a composite image of a human face superimposed on the digital animation.

The programme also seems to have tried to borrow another feature typically associated with New Media forms, namely non-linear narrative. This is done by splitting the major part of the documentary into seven fragments, each of which focuses on a different aspect of the production of the series. Every episode is clearly delineated with a separate title and opening sequence and although they are all related to the general topic of the series, each can function as a self-contained information unit, like entries on a CD encyclopaedia. The opening and closing sections of the documentary are largely made up of a montage of sequences taken from the series, that might be referred to as edited highlights inviting viewers to marvel at the visual spectacle that constituted a major part of the appeal of *Walking with Dinosaurs*. However, this attempt at remediation ultimately fails because the documentary's voice-over effectively functions as means of anchoring these fragments, imposing a linear sequential structure on them and guiding viewers along a particular narrative pathway. This highlights perhaps what is currently one of the crucial differences between the medium of conventional broadcast television and that of the Internet: the potential possibilities for interaction which they are able to offer the individual. Interactivity with television has until recently been limited to live studio debates, phone-ins or letters. Although new technological develop-ments will allow viewers to chose camera-angles in sporting events, to email programmes live or to Webcast, making live appearances on television via Webcam, this interaction is still very limited in comparison to the World Wide Web users' ability to navigate independently. We will return to this issue in our discussion of the Web site which accompanied the television programmes.

11.6 *Walking with Dinosaurs* – The Web Site

The official BBC Online *Walking with Dinosaurs* Web site became active in mid-September 1999, and amongst the many and varied offerings for site visitors was the chance to view a trailer for the series, prior to its television screening, still a relatively new concept at that time.[17] The Web site is still available

---

17   The trailer made obvious allusions to Spielberg's *Jurassic Park*, with its soundtrack reminiscent of the scene in which the thunderous footsteps of the gigantic tyrannosaurus are first registered as ripples in a glass of water and then heard approaching.

(February 2002) and currently the BBC has no plans to close down it down, as it is still visited regularly. Its continuing popularity is perhaps not surprising, since like many of the Web sites designed to accompany the BBC's flag-ship programmes, *Walking with Dinosaurs* provides a range of resources which are intended to be educational, informative and entertaining. The variety of formats in which these resources are presented provides an idea of the potential which the Web has to integrate and absorb all other more traditional media. Written texts, graphics, icons, images (including still photographs and video clips) and sound are used to present an eclectic mix of detailed facts and information about prehistoric life, glossaries of terminology, extracts from the *Walking with Dinosaurs* series, jokes, children's paintings and interactive games and puzzles. Hyperlinks connect to other sites likely to be of interest to visitors including a related notice-board and perhaps, inevitably, the BBC Online Shop which carries a full range of *Walking with Dinosaurs* merchandising, covering everything from books to cuddly toys.

Like many other media organisations in the 1990s, the BBC developed its Web site in order to complement its more traditional media products and although it is immensely popular with many different sectors of the audience, this new medium is still conceived of as a support for the Corporation's television and radio broadcasting rather than a rival. Perhaps it would be more accurate to speak of the *Walking with Dinosaurs* Web site as an interactive supplement to the television series rather than wholly a remediation of it. For although the Web site does indeed contain some of the images from the original series, these make up only a very small percentage of the network of resources on offer, many of which expand upon aspects of prehistoric life only touched upon in the television series. Indeed, visitors to the Web site may be initially surprised by the fact that a significant proportion of the material it contains is presented in the form of written texts which at first sight closely resemble book pages or articles from journals. However, the crucial differences between the traditional print medium and this kind of remediated text were pointed out by Ted Nelson, one of the originators of hypertext:

> Remember the analogy between text and water. Water flows freely, ice does not. The free-flowing, live documents on the network are subject to constant new use and linkage, and those new links continually become interactively available. Any detached copy someone keeps is frozen and dead, lacking access to the new linkage (cited in Landow (1992, p. 59)).

It is, then, the fact that resources on World Wide Web are refashioned to function as a vast interconnecting network which proves to be the most radical aspect of this form of remediation. For as Heinz Pagels has argued, "A network has no 'top' or 'bottom'. Rather it is a plurality of connections that increase the possible interactions between the components of the network" (cited in Landow (1992, p. 25)). This leaves the Web site user free "to choose his or her own centre of investigation and experience" and "not locked into any kind of particular organisation or hierarchy" (Landow, 1992, p. 13).

Curiously, given that much was made in the BBC Press pack of the use of the state-of-the-art digital technology in the production of *Walking with Dinosaurs*, the images taken from the series (whether still or moving) prove to be one of the least memorable features of the Web site. For whereas in the televisual context of the codes and conventions of wildlife documentary, these digital creations gave an impression of spectacular immediacy, once embedded as mere fragments in the hypermediated Web site their visual impact becomes minimal. Ironically, it is only when they are repurposed in a way that emphasises their status as hypermediated representations, for example when they appear as characters in the computer games or as brightly coloured cartoon-like icons indicating hyperlinks, that they succeed in capturing our attention.

We will conclude this discussion of the Web site by briefly considering another of its features which raises a number of interesting issues relating to immediacy, hypermediacy and remediation. Bolter and Grusin have commented that despite the seeming lack of significance of this kind of technology, "Web cameras are in fact deeply revealing of the nature of the Web as a remediator" (Bolter and Grusin, 1999, p. 204), and certainly this in true in the case under discussion. One of the Web site hyperlinks leads to a page showing the images produced by a Web camera, apparently trained on some exotic location. In the Web camera window, silhouetted against a bright orange sky – it appears to be sunrise or sunset – we can just about make out in the distance the head and neck of some huge prehistoric creature stretching up out of a forest of trees. This, then, can be considered as a particularly intriguing example of what Darley refers to as "impossible photography" in which "the computer has been used to produce the effect of photo-realistic representation in a scene that is conceptually fantastic in character – a scene that could have no direct correlate in real life" (Darley, 2000, p. 108). For although Web camera technology normally suggests transparent immediacy, offering Internet users an unedited stream of images of some location in the physical world, what we have here is hypermediacy. This Web camera simulation which reveals our fascination with media is the part of the Web site that most closely resembles the "double-take" or oscillating effect produced by the television images of the spitting tyrannosaurus in *Walking with Dinosaurs* or the sequence of the crew filming on location seen at the start of *The Making of Walking with Dinosaurs*. Yet again, transparent immediacy passes into hypermediacy before our very eyes.

11.7 Conclusion

As Bolter and Grusin have argued, the relationship between the older media forms (such as television, film and the printed page) and the New Media forms (such as computer animation and the World Wide Web) is complex and constantly evolving. A case study such as this one sheds light on the various ways that the different media find of interacting with each other and of refashioning representation, and also on the shifting balance between immediacy and hypermediacy which this process of remediation reveals.

Writing in the 1980s, Fredric Jameson voiced his fears about the consequences of what he called "the disappearance of the historical referent", which would leave us "condemned to seek History by way of our Pop images and simulacra of that history, which itself remains forever out of reach" (cited in Darley (2000, p. 72)). In his recent work about visual digital culture, Darley concluded that we have indeed reached a point when "computer imaging looks not so much to the world itself, as to already existing techniques of mediation, together with their attendant forms and styles. Prior forms, genres and works constitute a referential basis or ground for copying, acts of manipulation and recombination, and efforts aimed at further "perfecting" and simulating the already mediated" (Darley, 2000, p. 75). Thus, analysing the way in which we choose to represent dinosaurs, an extinct prehistoric life form, has much to tell us about the role which the New Media might play not only in shaping our understanding of the past but in determining our future relationship to reality itself in the postmodern world. As is often the case, to paraphrase the words of cultural theorist John Fiske, we take popular texts like *Walking with Dinosaurs* for granted when we should be taking them to pieces....

References

BBC (1999) *Walking with Dinosaurs*
Episode 1: New Blood
Episode 2: Time of the Titans
Episode 3: Cruel Sea
Episode 4: Giant of the Skies
Episode 5: Spirits of the Ice Forest
Episode 6: Death of a Dynasty
BBC (1999) *The Making of "Walking with Dinosaurs"*
BBC Online *Walking with Dinosaurs* Web site: http://www.bbc.co.uk/dinosaurs/

Adams, T. (1999) Television: the lost world. *The Observer*, 10 October, p. 16.
Banks-Smith, N. (1999a) Roar of approval. *The Guardian*, 5 October, p. 22.
Banks-Smith, N. (1999b) Last night's TV: I know thee, old man. *The Guardian*, 12 October, p. 22.
Barrett, P. (1999) A bone to pick. *The Guardian*, 11 October, p. 8.
Berger, J. (1980) *About Looking*. Writers and Readers, London.
Bolter, J. D. and Grusin, R. (1999) *Remediation: Understanding New Media*. MIT Press, Cambridge, MA.
Boucher, C. (1999) Review. *The Observer*, 3 October, p. 20.
BBC (1999) *Walking with Dinosaurs* Press pack, issued 10 September.
Brown, M. (1999) Kids' stuff? *The Guardian*, 13 December, p. 6.
Darley, A. (2000) *Visual Digital Culture: Surface Play and Spectacle in New Media Genres*. Routledge, London.
Gibson, J. (1999a) Dinosaurs give BBC1 much needed boost. *The Guardian*, 6 October, p. 9.
Gibson, J. (1999b) Why dinosaurs won't go away: how liopleurodon and his friends saved the BBC. *The Guardian*, 22 October, p. 2.
The Guardian (1999a) *Walking With Dinosaurs*, 4 October, p. 24.
The Guardian (1999b) Pass Notes: No. 1496: Dinosaurs, 7 October, p. 2.
The Guardian (1999c) Dinosaur science series "monster hit" for BBC, 20 October, p. 9.
Hanks, R. (1999) Television review. *The Independent*, 5 October, p. 16.
Horrie, C. (1999) An embarrassment of riches for the BBC. *The Independent on Sunday*, 5 September.
Jury, L. (1999) The future of the BBC. *The Independent on Sunday*, 8 August, p. 26.
Kellaway, K. (1999) Review. *The Observer*, 3 October, p. 16.
Kilborn, R. and Izod, J. (1997) *An Introduction to Television Documentary: Confronting Reality*. Manchester University Press.

Landow, G. P. (1992) *Hypertext: The Convergence of Contemporary Critical Theory and Technology.* The John Hopkins University Press, Baltimore.

Lawson, M. (1999) In Spielberg's footsteps. *The Guardian*, 4 October, p. 17.

Leiss, W., Kline, S. and Jhally, S. (1990) *Social Communication in Advertising: Persons, Products and Images of Well-Being*, 2nd edn. Routledge, Toronto.

Matthews, R. (1999) Why dinosaurs won't go away. *The Guardian*, 22 October, p. 2.

McCann, P. (1999) Rattled BBC chief attacks rivals for poaching formats. *The Independent*, 11 August, p. 4.

McKie, R. (1999) Who put the pee in the postosuchus? *The Observer*, 10 October, p. 11.

Peretti, J. (1999) A high-tech Sooty show. *The Guardian*, 7 October, p. 22.

Radio Times (1999) Jurassic parklife. 2–8 October.

Radio Times: Behind the Scenes Walking with Dinosaurs, October/November 1999.

Robins, J. (1999) BBC: barren, banal and confused. *The Independent on Sunday*, 15 August.

Steel, M. (1999) Yes, even I know my triceratops from my stegosaurus. *The Independent*, 12 October, p. 5.

Viner, B. (1999) Babes without the wood. *The Independent on Sunday*, 14 November.

White, A. M. (1999) To be blamed: the press in Britain. *TelevIZIon The Teletubbies*, 12 February, pp. 15–19.

Williams, V. (1999) *New Natural History*, National Museum of Photography, Film and Television, Bradford.

About the Authors

Karen Scott is based in the Education Department at NMPFT and also lectures in Film and Media Studies at University of Bradford. She is currently involved in doctoral research on the representation of the Past in contemporary television documentary.

Dr Anne M. White is Director of MA courses in Media and Cultural Studies, Department of Modern Languages, University of Bradford.

12

Trashing the Net: Subcultural Practice Online

Mark Goodall

Abstract

This intention of this chapter is to critically examine uses of the World Wide Web by fans of cult movies. It begins by outlining how cult movies are categorised, and notes the problems that this engenders. Then the relationship between technologies and subcultural practices is observed. Examples are presented to illustrate the question of whether, through remediation processes, such practices tell us anything new about forms of contemporary communication and consumption.

12.1 Introduction

> New media... blur the lines between film and fiction, reader and author, spectator and participant as well as between mass and elite culture. Such a democratisation of culture is as threatening as the widening of access to higher education to the traditionalists in education and the academy. (Cartmell *et al.*, 1997, p. 2)

Recent figures analysing current usage of new technologies confirm an increased take-up of the Internet, with the UK second only to the USA in level of enthusiasm for this new form of mass media communication tool.[1] The principal use of the Internet occurs via the World Wide Web – it's easy to use universally readable data processing and display via HTML and the HTTP protocol have encouraged ever more widespread production, dissemination and consumption of an often bewildering range of ideas, materials, images, sounds and texts. Some researchers have admitted technological developments are moving at such a pace that any serious study of its workings and complex relationships becomes dated and inaccurate very quickly (Gauntlett, 2000, p. 31). At the same time concerns have been raised about the increasing reliance on the Web itself as a tool for research. The popular Internet Movie Database (http://www.imdb.com/), for example, was recently embroiled in a dispute

---

1   Guardian/ICM poll: Techno Britain. *The Guardian*, 24 January 2001.

with a film director about the posting of inaccurate information about him on their site.[2] As much of the data submitted to Web pages is the opinion of unscreened individual enthusiasts, accuracy, in any traditional sense, often cannot be guaranteed. Thus the open access nature of the Web is perceived to have negative as well as positive implications. Despite this, critics and researchers cannot ignore the practices and activities of Web users, so this chapter hopes to further the understanding of how a specific strand of consumers feeds its unique pleasures online.

In this chapter I intend to look at the ways in which "cult" films are currently discussed, categorised and consumed on the Web. It is through observing how minority groups, with marginal areas of interest, have taken up the Web as a tool of mass communication that we can both indicate the breadth of usage of the Web and determine whether suggestions that the Web is a more advanced and convincing libertarian, communal, inclusive medium have any concrete basis. I intend to compare and contrast cult movie Web sites with those dedicated to more mainstream or "classic" forms of film art; to examine the structures which attribute to these films such a marginal status; and to illustrate how the Web is being used as a tool for communicating ideas and interests about "other" and "different" films to a diverse audience. Can the Web deliver more than previous modes of communication managed to consumers of cult product?

This chapter, however, is not – indeed cannot – be about cult movies *en masse*. There are, unhelpfully, as many definitions of what constitutes a cult film as there are books, magazines and Web pages about cult films. The ways in which cultural practices are categorised and defined have been articulated in research which has tended to highlight how attempts at labelling cultural products leads to a "canonisation" of particular works by certain elite groups in contemporary capitalist societies – media corporations, politicians, academic institutions, even legal authorities. It was the French sociologist Pierre Bourdieu who first developed a coherent study of how (different) taste defines and classifies groups of people in contemporary western society. Bourdieu argued that elite groups construct hierarchies of taste and maintain this through rejecting materials that are seen to be marginal in favour of a self-defined legitimate culture. His famous assertion linking politics and aesthetics was that:

> Taste classifies, and it classifies the classifier. Social subjects, classified by their classifications, distinguish themselves by the distinctions they make, between the beautiful and the ugly, the distinguished and the vulgar, in which their position in the objective classifications is exposed or betrayed. (Bourdieu, 1984, p. 6)

Bourdieu argued that this forming of legitimate and illegitimate cultural practices worked on the level of aesthetics. Bourdieu analysed French bourgeois culture, but in a popular cultural form such as cinema a distinct set of tastes and rules by which film materials are to be judged are established. Difference here (i.e. deviation from the "norm") is banished to the margins of theoretical

---

2   *Sight and Sound* (February 2001) reported that director Andrew Birkin had complained about misattributions on the site. Birkin also noted that none of the IMDB listings for the eight members of his family working in the film industry were correct.

interest or left to emerge in non-academic texts in a non-critical way. And the film industry itself sets up canonical hierarchies through fatuous awards ceremonies, promotional, non-critical articles in film publications, lists of the "best" films (arranged according to categorisations such as country/region, genre, historical period etc.) that advance certain, usually expensive, films as good. Bourdieu also noted that:

> Tastes are perhaps first and foremost distastes, disgust provoked by horror of visceral intolerance ("sick-making") of the tastes of others. (Bourdieu, 1984, p. 56)

Bourdieu's book was an expansive and considered account of the relationships between classes, tastes and power. But more recently theorists have noted how the phenomenon of cult/trash movies has become increasingly enmeshed with intellectual trends such as postmodernism where cultural relativism and celebration of difference produces new canons:

> The notion of cult effectively collapses the categories of "art" and "exploitation", eliding issues of both politics and aesthetics. (Cartmell et al., 1997, p. 86)

So what are notable here today are the ways in which material culture (in this case film) shifts between different classificatory systems and hierarchies of taste. Indeed, one of the key roles that academic study itself plays in this changing structure is to constantly reinvent and redefine new paradigms and extend boundaries of acceptability (Sconce, 1995; Chibnall in Cartmell et al., 1997; Thornton, 1995; Gripsrud, 1989). As a recent international conference on the subject of defining cult films clearly demonstrated, "marginal" media do not escape this complicated and contradictory socio-cultural relationship. Many of the papers given at the conference emphasised a "postmodern" convergence of those groups known as fans, academics and critics so that it was now possible to publish detailed academic studies of films hitherto seen as worthless of any serious interest and still be taken seriously – albeit as a "para-academic" (Cartmell et al., 1997, p. 81). Ironically, subcultural practices – special activities, interests collected around marginal cultural forms which at one time produce a high level of distaste amongst such guardians of cultural and social norms as listed above – become a part of acceptable society via those very institutions which once shunned them as unworthy.[3] Those clearly delineated boundaries of "academic/researcher/archivist" and "fan/collector" are arguably collapsing or, more interestingly, *merging*. The bond that develops between subcultural activity and emerging media technologies aids this further. The usual pattern is for subcultural groups to find themselves (and their uses of media) scrutinized by the dominant media forms, from which new hybrid uses of media technologies are then developed as counter-attacks by the subcultural group.

As this chapter is paying particular attention to the role of the Web in these new relationships it is useful to note the very recent attempts at reworking how such

---

3   Thornton (1995, p. 11) observes that: "Subcultural capital can be objectified or embodied... objectified in the form of fashionable haircuts and well assembled record collections... embodied in the form of being 'in the know'".

new media are studied, researched and observed. Gauntlett (2000) outlines some useful ways of analysing new media, particularly the Internet. Gauntlett notes how at first glance the Web and film seem to be opposite media – Web pages in fact share more in common with magazines. He then goes on to identity three basic uses of film on the Net: publicity, review and production. These will be the consideration of this chapter.

12.2 What Are Cult Movies?

> A cult object... must provide a completely furnished world so that its fans can quote characters and episodes as if they were aspects of the fans' private sectarian world... the adepts of the sect recognise through each other a shared experience. (Eco, 1986, p. 198)

It is certainly safe to say that the Web has been central to the further development of interest in cult media. Whilst this has been focussed on particular TV programmes (*Star Trek, The X-Files, Xena: Warrior Princess*), film is rapidly becoming part of this discourse. A very famous example of how a film became identified as part of a "cult" was *The Blair Witch Project* (1999), whose success and mystique came about due to strategically placed promotional devices via Web sites (such as http://www.blairwitch.com/). Obsessive interest in the film, before it was commercially released, thus became self-generating and the cult of the "Blair Witch" still lives on.

I would like to proceed in time to a specific case study to illustrate this. However, it is first necessary to note some of the various requirements that must be fulfilled in order that a film can be defined as belonging to some sort of "cult". Key to this is that many cult films are so defined because of their (often difficult) relationship with the various agencies of control that define, shape and authorise the status of films (film and television studios, video companies, distributors, academics, critics, as above), so the following are important factors:

- Film "style"
  Many current debates around cult movies are centred on so-called "Badfilm" (Sconce, 1995, p. 371). These are films which have for some time been critically lambasted as badly made, with poor acting, weak scripts, unimpressive special effects and incongruous editing which appear to undermine most of the "rules" about film art. Despite this, a specialist following develops. Examples of these can be individual films (Paul Verhoeven's *Showgirls*, 1995) or the entire output of a single director (Ed Wood, for example).

- Problems of access
  Certain films enjoy cult status because they are obscure and/or have been forgotten by the various agencies of control, or are "foreign", avant garde or "difficult" in terms of style and content.[4] Whilst the advent of domestic video

---

4   In their book *Cult Movies*, the French brothers for example list Alain Resnais' art-house obscure, and "difficult" *L'Annee Derniere a Marienbad* (*Last Year at Marienbad*, 1961) as a cult film.

extended the availability of many "lost" films they were not immune to being discontinued/deleted and thus acquiring further "collector" status.

- **Repetition and mimicry**
Some films become the subject of intense loyalty as agents of social gathering and appreciation. Films such as *The Rocky Horror Picture Show* (1975) and *The Blues Brothers* (1980) enjoy a large and devoted following of fans that are known to organise conferences, appreciation societies and fan clubs where fans of the film gather to dress up and recreate scenes from the film.
- **Fashion**
As cultures and societies "develop", certain films or film styles and film narratives and representations become prescribed as "outdated". This may be built around political change and attitudes to sexuality, gender and race, and films that have ignored or been outside of such ideological shifts are noted to be of a reactionary nature to what becomes socially acceptable. Such films may however be later reconstituted as camp or kitsch (British "sexploitation" film for example).
- **"Cult" becoming "Classic"**
It has been noted that media products can enjoy transference of cultural capital and films are no exception. For example, British films such as the gangster movie *Get Carter* (1971) and supernatural horrors like *The Witchfinder General* (1968) or *The Wicker Man* (1973) once languished in the bins of their production and distribution companies (in the case of the latter quite literally). But due to various cultural and social changes, engendered by critical re-readings and certain cult audiences, such films have seen their stock raised to that of classic. These films, which were rarely issued on video, now regularly feature in "best of" lists and have recently found themselves being remastered and issued on domestic formats. Likewise, Americans Russ Meyer and Roger Corman, and Italians Dario Argento and Mario Bava have been, or are in the process of being, rediscovered as "great" directors.

12.3 Technology and "Cultdom"

There is nothing new in the relationship between emerging technologies and the consumption of subcultural media. In the arena of film each new technological development encourages new forms of cultural consumption. In the 1960s film critics began to make their own films that often paid tribute to films of the past.[5] The advent of television eventually saw a need for the filling of extended broadcasting – this merely increased with the satellite and cable revolution in the mid-1990s. For this purpose cult and exploitation films were unearthed and are still being shown (Channel 4, Film Four). Then, when home video exploded in the early 1980s, a need again developed for the filling of rental house shelves with product. Many "paracinematic" (Sconce, 1995, p. 372) genres took on new

---

5 Or "movie buffs" as the French brothers define them (French and French, 1999, p. 7). The authors are referring to the directors of the French Nouvelle Vague (New Wave) and young American filmmakers influenced by them, such as Peter Bogdanovich.

leases of life here. Once consumers were able to record films directly from TV broadcasts, collections of filmic material expanded yet further. Furthermore, the ability to make multiple copies of taped films helped spread the consumption of cult film and engendered secret societies of fans through magazines and fanzines. The recent development of DVDs has encouraged film companies to delve further into their archives for product. It is conceivable that a revolution similar to that of the music Compact Disc, where an immense amount of material has been reissued, re-mastered even re-compiled, will take place. Certain forms of cult movie have quite bizarrely been unearthed for the digital revolution.

The case of print publications is worth special note here, as this is the medium with which most consumers of cult film have engaged as a means of dissemination, communication, distribution and trade. Magazines and fanzines appear to have taken a back seat to these newer interactive media; certainly they have a lower profile in studies of media now.[6] Yet they are still hugely important for the consumption of cult movies. Cheaper DTP technology and printing and reproduction costs have improved the quality of fan magazines. Yet the visual style of such fanzines harks back to a deliberate aesthetic of low quality that determined and set apart the fanzine from its more glossy relations. There is no attempt at pleasing a broad audience – these publications are for those in the know. That said, niche marketing is a growth area (as any browse through the immense racks of bookstores like Borders can demonstrate). Magazines devoted to cult movies are simply another niche to be catered for.

Bolter and Grusin (1999) developed the idea of "remediation": a theory of how new media, far from replacing old media, actually rely on their stylistic and functional traits to rework new styles. We must therefore examine the remediation of magazine into Web pages. Cult film magazines, like their Web counterparts, are often edited if not written by a single individual who becomes synonymous with their publication.[7] Web sites, like their print cousins, often have a section devoted to the creator that tends to be biographical but also can act as a manifesto outlining their particular tastes and definitions of what constitutes a "cult". So the deployment of the Internet as a tool for the furthering of cult movies can be identified as part of a familiar relationship between consumption of film and emerging technologies.

The importance of being able to trade and exchange film product is still key to the cult movie subculture: now, as then, the ease with which tapes can be copied can be evidenced online. The fact that low-budget exploitation films in particular are rarely encoded with expensive copyright disablers, or are usually out of

---

6   Fanzines are described by Sconce (1995, p. 372n) as "home-produced photocopied magazines circulated amongst fans and devoted to an often narrow area of interest in popular culture".

7   The best illustrations of this aesthetic are Michael Weldon's *Psychotronic Video* and Bill Landis' *Sleazoid Express*. Both these editors are absolutely devoted to the cult film and its ethics with low-budget production values strongly evident in their publications (*Sleazoid Express* is still photocopied and stapled together and has no Web or email address).

copyright, has helped the exchange immensely. (DVDs, however, are another matter.) So-called cult or underground films that had all but disappeared from the theatrical circuit[8] were given new leases of life by this inexpensive and simple means of exhibition.

12.4 Mondo-net

Finally we come to some examples of cult Web sites and particular case studies. In searching the Web for information about such films I discovered some interesting uses of the Web for the collection and dissemination of cult films that may have useful implications for other such specialist interests.

Traditionally differences between big budget and low/no budget films are clearly reflected in printed media. This results mainly from ideas about how "good" and "bad" film publications are constructed. Stereotypically it is assumed that Web sites devoted to minority or cult films are produced by an obsessive loner desperate to share their devotion with other similar social inadequates. If it has been noted how:

> It is striking that the popular stereotype of the nerdish fan – "anal", socially inept and badly dressed – neatly coincides with that of the academic. (Cartmell *et al.*, 1997, p. 10n)

then we can see how the common stereotype of the "computer geek" (that mythical group who make particular use of computer technologies) can be added to this list. Yet even the most official and well-respected film sites began life as the obsession of an individual. The Internet's most popular film site, the Internet Movie Database, began life as a personal project before being sold to US giant Amazon when the link was made between detailed information about particular films and the possibility of commercial gain by selling copies of those same films via trading sites. A more careful examination of some cult movie Web sites, however, problematises this cosily clichéd world view and suggests that remediation of representations is developing new possibilities and applications.

It is noticeable at the present time how Websites that are concerned with the exploration of what may be deemed cult movies can be divided into a number of functions. The principal functions are discussed below.

12.4.1 Archival

Websites promoting in part or as a whole, the collection and presentation of films that have been ignored or forgotten by the dominant cultural structures appear to have been established. This activity can counteract the combination of ignorance and over-simplistic categorisation of films as marginal either as dated or built around subject matter seen to be of little historical interest. If one were to

---

8   Except in famous repertory cinemas such as London's La Scala and New York's Cameo Theatre.

Figure 12.1 Ident for Something Weird Video.

identify the sorts of films that have received archiving and/or preservation and replication (by the British Film Institute for example), there will be very few marginal films on this list. The BFI's commissioned list of 100 best films of the century featured mainly films which are considered by canon to be "classic".

The best illustrative example of this is the American company Something Weird Video (SWV; see Figure 12.1) (http://www.somethingweird.com/), which since 1990 has collated and "saved" a vast range of forgotten film product from the history of the moving image. Although Something Weird is a commercial operation and therefore is not bound by methodologies of existing archival archaeological work, they are nevertheless acting as preservers of late twentieth century cultural ephemera. As well as rescuing forgotten feature and short length films from the "dustbins" of film distribution and library sources, together with those of the private collector, SWV has produced compilations for example of instructive films – taboo subject matter which acted as "educators" of American youth in the late 1940s, 1950s and 1960s.

The creator of SWV is Mike Vraney, and his philosophy is outlined on the site:

> Here on your screen is a whole world of film that just a few short years ago was considered lost or worthless. The industry that produced and distributed these films had long since vanished and there was no sign of the men who actually created these bottom of the barrel celluloid wonders....
>
> In my mind, the last great genre to be scavenged were the exploitation/ sexploitation films of the 30's through the 70's. After looking into this further, I realised that there were nearly 2,000 movies out there yet to be discovered. ...The future looks bright for us with over 1,500 films left to find, cable and television opportunities, CD-ROMs, and all the other endless possibilities out there... (Something Weird online editorial)

These films may appear very marginal, but will probably act as important documents of post-war culture in years to come and a valuable tool for research into responses towards post-war youth culture. In an age obsessed with the new and hyperreal, SWV has employed new and cheap transfer technologies to go back and recover lost materials for posterity. Even SWV's collections of pornographic loops, whilst undoubtedly voyeuristic and titillating, reflect ideas about sexuality that as soon as they become laughable or even "quaint" will still need to be archived and studied.

The Internet is an increasingly important method for companies like SWV to make money from the sale of these films and thus secure the continuation of their business. A global audience can once again enjoy their bizarre collection of "low" film oddities in high-quality analogue and digital.

12.4.2 Commercial

Something Weird claims to adhere to high production values for their products. But many Web sites devoted to "cult" films act as providers and sellers of an even vaster array of product with a devotion to customer service that would put many bigger film and video retailers to shame.

The importance of online trading for purveyors of cult film material is clear. Companies like Luminous Film and Video Wurks (sic) (LFVW; http:// www.lfvw.com/) and Trash Palace (http://www.trashpalace.com/) estimate an 80% share of business coming from online customers.[9]

While streamed video of trailers or tasters of film product are well established by sites such as IMDB, cult movie sites are no strangers to the capabilities of Web interaction. Fred Frey's LFVW presents a bewilderingly complex array of material for the cult movie fan, including secure order forms/baskets, film clips, pictures, multimedia, surveys, posters and even Real Audio sound files to be enjoyed whilst perusing the site (see Figure 12.2). The construction of promotion for some of the films on offer is incredibly sophisticated, linking film stills, artwork and synopsis together with advertising links all which belie the fact that LFVW is a one-man show.[10]

A promo for the notorious exploitation work of "Joe D'Amato" demonstrates the capabilities of inventive text and graphics and is arguably a more dramatic use of Web capabilities than the presentation of more "conventional" film product.[11] Rieder (Gauntlett, 2000, pp. 96–97), like others, has lamented the poor design of many Web pages as cluttered, gimmicky and cumbersome to use. Despite its apparently dubious content, LFVW makes bold and simple use of graphics and text to promote its wares. The urban myth that the most effective Web sites in terms of downloads, access and speed through attention-grabbing frames, banners and links are porn sites is here recalled. If, as Bolter and Grusin (1999) have argued, Web pages are hypermedia, then film sites tend towards

Figure 12.2 Banner advertising *Avere Vent'anni*.

---

9   Statistics from responses to an email questionnaire sent out by the author to various cult movie Web sites.
10   The example here is Fernando Di Leo's notorious multi-genre cult *Avere Vent'anni* (*Being Twenty*, 1979), which is heavily promoted on the LFVW site.
11   D'Amato's real name is Aristide Massaccesi.

presenting the tricks that HTML, streamed video and graphics can offer. There is little point in Web sites trying to emulate the scale of the film experience – digitised video images on Web pages are even more inferior than home video in terms of quality. What they can do is emulate the medium of print, much as the sites for the magazines Cult Movies (http://www.cult-movies.com/) and Shock Cinema (http://members.aol.com/shockcin/main.html) do. The Italian cult film magazine *Nocturno*, although glossier than its US counterparts, has a Web site that features similar articles from the magazine. If, as noted above, fanzines are laid out and constructed to "appear cult", then Web pages emulate that also. The large proportion of space in cult movie magazines devoted to advertising and personalised articles written by fans of films, rather than professional "critics" or "academics", can be mirrored in the Web pages. What the Web pages open up to is a more global audience – non-native consumers of cult videos are unlikely to write or phone orders through, but will more happily access product online. This is particularly the case in the UK, given that many domestic VHS machines and some DVD players can read NTSC tapes/Region 1 DVDs.

12.4.3 Community

The act of bringing together individuals with similarly eccentric and devoted obsessions common to discussions of fandom and cult becomes a global (if virtual) activity with the advent of digital technologies. Fans of cult movies can exchange personal information about tastes, photographs and images in the form of graphics files, and can swap and exchange video tapes, CD-Rs, text pages and scans of paper articles and names and addresses (both email and snail mail) with relative ease. The ability that the Web has of publishing and promoting global events – festivals, fairs, conferences – is utilised fully here as a means of expanding, paradoxically, the audiences for cult movies (even if in reality this does not happen; i.e. not may UK fans travel to conferences in the USA).

An interesting aspect of such communal sharing of information through Web sites is the links, which can often reveal a bewildering array of "related" sites, many of which often have very little in relation to the subject site. It must be noted, however, that it is in the act of creating links to other sites the author of the host site develops coherence between notably different areas of the filmmaking business. Dominant Web sites such as the IMDB may even be creating new canons, a process that has strong historical precedents in the fanzine area (Sconce, 1995).

The Internet helps bring together easily, if virtually, individuals that would otherwise have difficulty communicating efficiently about their interests in cult movies. Things can happen very much more quickly and more cheaply than they would do with more conventional communication tools (phones, letters etc.). When fans online access and exchange photographs of film events it may seem to be obsessive behaviour to outsiders, yet these are the kinds of activity that bond these subcultural communities together. US critic Stanley Fish defined this interaction as "Interpretive Communities", where:

...groups congregate around the meaning made from the interpretation of various texts. (Cartmell *et al.*, 1997, p. 6)

The genre we are now going to focus on fits into the "dated" category. Now residing within the field of the exploitation film "Mondo" films were sensationalist documentaries (or shockumentaries) that set out to reveal the oddities of the world's cultures – those developed and developing alike. Originating in Italy in the 1960s with Gualtiero Jacopetti and Franco Prosperi's *Mondo Cane* (1962), from where the genre title was generated, variations on the theme were made in the USA, Britain, Germany and Japan (where Mondo films were enthusiastically consumed).

The Internet Mondo Movie Database (IMMDB; http://utenti.tripod.it/ immdb/) is an Italian Site hosted by Luca Persiani and devoted to Mondo films. Clearly identified as a "fans'" Website, Persiani's site bears all the hallmarks of a cult site. Not only do the disclaimers and editorial comments mark out the specialist and controversial nature of the site, but the site itself is hosted by an organisation called "1000 e non piu' 1000 (1000 and not more than 1000): Apocalypse Culture Webring" (http://members.tripod.com/~StefaniaD/ mille.html), whose elite philosophy is summarised through a series of requisites: anti-political correctness, devotion to those "Apocalypse Cultures" defined as serial killers, new cults, conspiracy theories, trash culture, strange music, terrible art and of course cult movies.[12] There then follow instructions and HTML fragments to help construct Web sites for the "ring" as well as membership forms (see Figure 12.3).

The dynamic of shock – key to the success of much Mondo filmmaking – is employed to maximise the interest in the site. As noted above, Web sites are capable of exhibiting all the key traits of hypermediacy – fragmentation, emphasis on performance rather than a finished art object. Yet many sites that are concerned with the Mondo genre pivot around the notion of voyeurism – a form of immediacy. Links from the IMMDB, for example, are concerned with film and video footage and photographs of "real" often traumatic events. Thus the visitor is transformed through hypermediacy to immediacy. A notorious

Figure 12.3 Logo for "1000e non piu' 1000".

---

12   This definition was developed by Adam Parfrey in his 1990 collection *Apocalypse Culture*, now into several volumes, which collects numerous extreme and disturbing cultural practices under the domain. Translations from Italian Web sites are the author's own.

piece of American TV footage recording the on-screen suicide of a US senator has become a widely circulated streamed piece of "real" footage.[13]

These links mimic the underground networks of film fans who would have resorted to word-of-mouth or film fairs to make fans of particular films cohesive. If it is the act of representing a film text, the particular ways with which a film is advertised, promoted and visualised rather than the content of the film itself that categorises it as "exploitation", then the Web sites mentioned above conform neatly to that practice. They are out to sell videos and DVDs, posters, records, stills etc., and they use every tool the Web has to offer to try to do this.

12.5 Conclusion

Web sites concerned with the specific film genres of cult, exploitation and by extension more specific types of film (Mondo) have remediated the format of the magazine/fanzine in order to promote and extend appreciation of obscure films amongst like-minded audiences. As with other Web-based information on films, hard and reliable information is difficult to come by. The principal success of Web pages devoted to cult movies is the widening of availability of copies of the films themselves; at present this is through the form of VHS video, although some success has been noted with the sale of DVDs of reissued exploitation/horror films. With their confused and overloaded hypermediated format, many large sites devoted to mainstream movies are confusing and unpleasant to navigate. Cult movie Web sites "remediate" the low-budget approach of the fanzine, yet many display all the hallmarks of the "hard sell". The studies of "fandom" that have emphasised "community" are evident in uses of computer-mediated communication: email, bulletin boards, chatrooms etc. (Pullen in Gauntlett (2000, pp. 52–61); Smit in Gauntlett (2000, pp. 130–136); Pearson in Cartmell *et al.* (1997, pp. 143–161). But this paper has been concerned with the commercial aspect of subcultural online practice. Here the key is good design, and as with previous modes of new technology those that are experimenting with the medium have most to gain in terms of visual impact.

The audience for cult movies is widening as more and more films from this subculture find their way into respectable academies, be these academic or commercial. Some of the pioneers of the study of cult films have already expressed an interest in moving away from this field into the study of middlebrow cultural forms (the upmarket tabloid newspapers, soft porn). By doing this they are confirming what critics have suspected all along: that once something becomes too popular it is no longer a cult.

---

13   The footage referred to here is news footage of the suicide of a Pennsylvanian State Treasurer accused of corruption, R. Budd Dwyer. Dwyer shot himself through the head in front of TV cameras.

References

Books

Bolter, J. D. and Grusin, R. (1999) *Remediation: Understanding New Media*. MIT Press, Boston, MA.

Bourdieu, P. (1984) *Distinction: A Social Critique of the Judgment of Taste*. Routledge, London.

Cartmell, D. *et al.* (eds.) (1997) *Trash Aesthetics: Popular Culture and its Audience*. Pluto Press, London.

Eco, U. (1987) *Travels in Hyperreality*. Picador, London.

French, K. and French, P. (1999) *Cult Movies*. Pavilion Books, London.

Gauntlett, D. (ed.) (2000) *Web.studies: Rewiring Media Studies for the Digital Age*. Arnold, London.

Gelder, K. and Thornton, S. (eds.) (1997) *The Subcultures Reader*. Routledge, London.

Kerekes, D. and Slater, D. (1995) *Killing for Culture: an Illustrated History of Death Film From Mondo to Snuff*. Creation Books, London.

Thornton, S. (1995) *Club Cultures: Music, Meaning and Subcultural Capital*. Polity, Cambridge.

Journal Articles

Gripsrud, J. (1989) "High culture" revisited. *Cultural Studies*, 3(2).

Sconce, J. (1995) "Trashing" the academy; taste, excess, and the emerging politics of cinematic style. *Screen* 36(4).

Magazines

Cult Movies (USA, ed. Michael Copner)
Nocturno Cinema (Italy, ed. Manlio Gomarasca, Davide Pulici)
Psychotronic Video (USA, ed. Michael Weldon)
Shock Cinema (USA, ed. Steve Puchalski)
Sleazoid Express (USA, ed. Bill Landis)
Video Watchdog (USA, ed. Tim Lucas)

Web Sites

Bad Cinema Diary (http://www.cathuria.com/bcd/)
Cult Movies and Movie Treasures (http://www.geocities.com/Hollywood/Cinema/5034/index.html)
DVD Cult (http://www.dvdcult.com/)
Exploited Films (http://www.exploitedfilms.com/)
Internet Mondo Movie Database (http://utenti.tripod/immdb/)
Internet Movie Database (http://www.imdb.com/)
Luminous Film and Video Wurks (http://www.lfvw.com/)
Mario Bava Web Page (http://www.mariobava.tripod.com/)
Mondo Culto (http://www.mondoculto.com/)
Mondo Digital (http://www.mondo-digital.com/)
Mondo Video Art (http://www.delerium1.demon.co.uk/delirium/m_titles.html)
New Media Studies (http://newmediastudies.com/)
Nocturno Cinema (http://www.nocturnocinema.com/)
Pimpadelic Wonderland (http://pimadelic-wonderland.negation.net/home.html)

Shock Cinema (http://members.aol.com/shockcin/main.html)
So Sweet So Perverse (http://so-sweet.cwc.net/index.html)
Something Weird (http://www.somethingweird.com/)
Trash Palace (http://www.trashpalace.com/)
Video Watchdog (http://www.cinemaWeb.com/videowd/)

13

Another Time, Another Place: The Use of Technology to Make the Cultural Heritage of the Organ Accessible

L. K. Comerford and P. J. Comerford

Abstract
This chapter discusses different ways to use technology to make available to serious organists and students of organ history the complex and subtle sounds illustrating different organ cultures. This includes the use of a sound synthesis technology designed specifically to take account of these requirements. It includes findings from experimental work aimed at identifying and meeting the particular challenges of this application.

13.1 Introduction

Varied styles of organ sound have developed in different national and regional cultures, and over time (see, for example, Sumner (1981), Wills (1984) and Baker (1991)). The developments have been influenced by varieties of organ use, national musical traditions, liturgical practice and architectural styles; these have influenced the positioning of the instruments and the relationship between the different departments of the organ. In combination, this has resulted in varied tonal development. For example, in mid-19th century England, liturgical changes transferred the choir to the front of the church, as a copy in miniature of the choral arrangement at cathedrals. As a result, nearly every organ in a parish church which had previously been sited in a back gallery was moved too, and came to be typically sited in a transept, in a side aisle or most often in a recessed chamber at the side of the chancel, fronted by a "fence" of pipes (either sounding pipes or dummy display pipes). This position, off the main building axis and often surrounded by masonry, meant that larger organs and higher wind pressures were needed to produce an adequate volume of sound. The increased size meant that the layout was often cramped, so that departments are blocked horizontally by one another. A horizontal layout, coupled with a surrounding chamber, produces an indirect sound with complex acoustic reflections; this is ideal for the production of soft Romantic tone colours, but means that sound tends to become more indistinct when heard at a distance than from an open plan organ – in particular, the higher frequencies are heavily attenuated. Classical organs on the other hand tend to be free-standing – often elevated – on the main building axis, rather than built into recesses; many follow the "Werk" principle, which means that the relative placing of different departments is vertical – in this way, each department can speak as an independent entity, unobscured by the others. Each department is housed in its own open-fronted case, with a roof. This arrangement throws sound energy forward without any impediment, so the pipes can produce an equivalent output of sound to that of an impeded organ, yet are blown less hard, so improving the quality of tone produced. The power of the classical organ is derived less from indirect sound than is that of the Romantic organ.

Although the organ is called the King of Instruments, it is in fact more like a whole dynasty with different royal lines, each with its own individual but related history.

Such different developments have given rise to music written especially with each style of instrument in mind. Although it is perfectly possible to execute these pieces on organs of alternative styles, just as it is possible to make piano transcriptions of pieces originally intended for orchestra, there are elements of the music which come alive in a unique way when heard on the type of instruments for which they were intended. For example, organs in the Romantic tradition have tremendous dynamic flexibility by virtue of having stops with a wide dynamic range (from very loud Tuba to very soft Vox Angelica), with different departments contrasting in power; the contrast is often further increased by enclosing some departments in a swell box from which the amount of sound egress through shutters is controlled by the player. This great difference in stop power means that is not possible to freely blend the tone colour of any stop with that of any other, as the loud will swamp the soft. The Romantic style of organ composition exploits these contrasting stop powers, and where applicable makes a virtue of the mellow tones resulting from enclosed pipes, where a high proportion of indirect sound is present. In the Classical organ, stops and departments tend toward a more common dynamic level, so it is therefore possible to play together almost any pair of stops or group of stops without one overwhelming the other, and this gives the potential for many matching tonal combinations, such as are exploited in fugal writing. Each regional tradition of music and each period has its own distinct features which reflect the characteristics of the instruments for which it was written.

In the visual arts it is arguably perhaps sufficient to record and catalogue such historical and regional variations, museum-style, for the information and enjoyment of future generations or of those from other traditions. With a musical instrument, however, merely to record the sound is useful, but it does not make the whole musical experience accessible to the player and musician as a living entity.

What is ideally needed is a way of reproducing the sound of each organ style, so that it can be played as a real instrument in any location today, and can be compared with the sounds of other historic or national flavours. In this way the rich heritage of organ tone can be kept alive, not just mummified, and such a facility can make different styles of instrument sound available to musicians and organ scholars who would not otherwise be able to experience them. The sounds have to be not just sampled, but adaptable, so that they can be interpreted and expressed through a different medium and, very importantly, in a different acoustic.

13.2 Accessibility of Cultural Heritage

Against this background, the challenge of accessibility can be approached in a number of ways.

The simplest approach is one of documentation rather than musical experience – to use the Internet to distribute the details of examples of different styles of organ (such as is achieved in the Osiris library (http://www.wu-wien.ac.at/earlym-l/organs/local.html) and by specialist organ sites), accompanied by details of recordings of their sound in performance (as in extensive organ discographies). This allows the organ student or enthusiast to build up a reference library of information and sounds. A slight refinement is to load the performance recordings directly from the Internet (brief illustrations of organ sound are often available on church Web sites, but are often of poor quality).

A much more imaginative approach is to create an instrument in the style of the original, which can be played by an organist far removed from its original location. In practice, to satisfy the serious organ student, the worth of this approach is constrained by the need to find a solution which takes account of the specialist requirements of organ tone by fulfilling the following criteria:

1. It must be powerful enough to handle the detailed structure of each organ sound, including the complex attack transients.
2. It must be capable of handling the complexity of a vast ensemble of simultaneous sound, which is one of the essential characteristics of organ tone.
3. It must be flexible enough to take in the varying requirements of different cultural developments.
4. It must be able to adapt the sound of an instrument originally installed in one acoustic so that as far as possible its character is preserved when heard in another acoustic.

Three basic alternatives are available, described below.

13.2.1 Sample Loading from the Internet

Samples of sounds taken from original instruments can be loaded from the Internet and edited to provide sounds playable from a midi keyboard. (For example, the Pipe Organ Samples Web site (http://home.t-online.de/home/andreas.sims/esamples.htm) offers this facility as a demonstration of a more extensive range of organ samples available on CD). From the point of view of the serious student of organ tone, this has a number of problems. One, to be discussed in more detail below, is that acoustic organ sounds are never "dumped" from one acoustic to another; they are always re-voiced to take account of the local acoustic conditions, which is not possible if sampled sound is used. In addition, the sampling process of recording, processing and play-back though loudspeakers introduces colorations which need to be compensated for if an accurate reproduction is to be achieved. (Loudspeaker selection is of particular importance for the output of organ sound (Comerford and Comerford, 1996), and is therefore an intrinsic problem if sound is to be broadcast over the Internet with no control over the characteristics of loudspeakers used.) A second major problem is the player interface. Although different organ traditions have developed a variety of console styles, with stop selection switches being drawstops, stop tabs or rocker switches, and arranged in a variety of characteristic ways, most have in common that the switches give a clear indication of which stops are currently selected, and that more than one manual keyboard is used, with a large-scale pedal keyboard played with the feet rather than the hands. These keyboards control the different departments of the organ, which can be played separately or coupled together. These player controls are an integral part of the playability of the organ, and are difficult to realise adequately using a non-specialist interface.

13.2.2 Real-time Playing from the Internet

An alternative to pre-loading samples is to play in real time over the internet – that is, select the switches and play the notes on a midi keyboard at this end, send that information to the remote receiver and receive back in waveform format the sounds appropriate to those switches and notes. This approach was suggested by Professor Rae Earnshaw of the University of Bradford.

To ensure good quality undistorted organ waveforms, the following features are necessary:

- *Adequate sample output rate*
  Note that the frequency range of an organ is uniquely wide amongst musical instruments, encompassing stops of different pitch, each with a 61-note range and with each note with a fundamental and a range of higher partials. The top note of a 1 foot organ stop on a five octave keyboard would need a sample output rate of 37 kHz or higher to play its 16.7 kHz fundamental and a sample output rate of 74 kHz or higher to play its 33.4 kHz second partial (inaudible to humans).

- *Sufficient bit resolution maintained over a sufficiently wide loudness range; adequate signal to noise ratio*
  Modern digital organs usually employ a minimum of 16 bits for storage and for waveform output. For waveform output it has to be remembered that 16 bits has to cover a wide dynamic range on each channel and, although loud sounds will use waveform outputs at or near full resolution, quieter sounds may be coded in considerably less than 16 bits. Reduction in the number of active bits reduces the signal to noise ratio by 6 dB per bit. Because of the very wide dynamic range of the organ, it is therefore necessary to have a mechanism to preserve at or near full waveform output resolution over a dynamic range of 48 dB or greater.
- *Low distortion, low coloration and low bandwidth loss due to analogue and audio components*
  Real-time playing from the Internet presupposes a transfer rate sufficiently fast to keep up with the supply of complex sound data required to create a musical performance. At present, the Internet transfer rates available to the average student of organ tone would in practice restrict the waveform data for the whole instrument to one or perhaps two channels of limited bandwidth, whereas one of the characteristic features of organ tone is the distributed and independent nature of its sound sources. These restrictions can be overcome at a cost, and once this option is more economically viable, this approach becomes more attractive to the organologist, and long term, if synthesised as opposed to sampled sound is received, it may be possible to adjust this remotely to suit the destination acoustic. The question of player interface remains.

13.2.3 Specialised Synthesis Technology

An alternative approach is to use a technology developed specifically to address the problems of organ tone synthesis. Although not offering the breadth of accessibility of an Internet solution, this embraces all the specialist features that an Internet solution will need for this application. The remainder of this chapter is concerned with describing the use of such a system (Bradford Enhanced Synthesis Technology: BEST) in the context of increasing accessibility to organ cultural history. In addition to incorporating the necessary features for production of undistorted waveforms, as set out under the previous heading, the development of BEST for this purpose required that the four specialist requirements noted above were addressed, as described below.

13.3 Critical Components of Organ Waveforms

As part of a study of organ sound and its synthesis funded by the European Commission, a series of experiments was undertaken to determine the factors in organ sound structure which were most critical in the perception of organ tone and in judgement of its quality. The study then went on to identify methods of ensuring that these factors were reproducible and adjustable with a synthesis technology such as BEST.

Through these experiments, detailed waveform content factors were identified as critical to the overall sound structure of various classes of stop. Key cross-cultural factors identified were:

- amplitude and frequency disorder of partials during attack and decay phases
- wind noise
- amplitude and frequency instability during sustain phase
- harmonicity/inharmonicity of partials
- tuning and pitch variations
- irregular note tonality

Culture-specific factors were examined further in the work on cultural variety described below.

13.4 Complex Simultaneous Sound

Although organs have developed differently in different cultures, particularly in tonal structure, one cross-cultural feature they have in common is the presence of multiple pipes in each stop, stops in each department and departments in each instrument. As each individual sound has the complex structure outlined in the previous section, this multiplicity of sound sources gives the potential for a great complexity of sound to be produced simultaneously.

Ensemble in this context is defined as the perceived size of sound-producing resources. While ensemble is to a degree linked to loudness, its level is primarily determined by the number of active independent sound producing sources or by what this number is perceived to be. For example, the difference between a string quintet and a full string orchestra playing the same musical parts is not simply the loudness level at a listener's ears, and if the string quintet is amplified to the loudness level of the full orchestra the same result will not be obtained. For a high degree of perceived ensemble, what appears to be needed is a multiplicity of sound sources distributed in space, not precisely in tune or in synchronism (but not too far out), and, perhaps, with a degree of dynamic variation in their loudness and tonality. The experimental results highlighted the differing characteristics of different families of stops, and of stops from different traditions.

13.5 Voicing Flexibility for Cultural Variety

Once the cross-cultural critical factors of organ tone have been identified, and mechanisms and tools for their synthesis have been developed, it is important to check that these tools can indeed produce sounds which will fulfil the brief of quality organ tone in a range of organ cultures. To achieve this, a

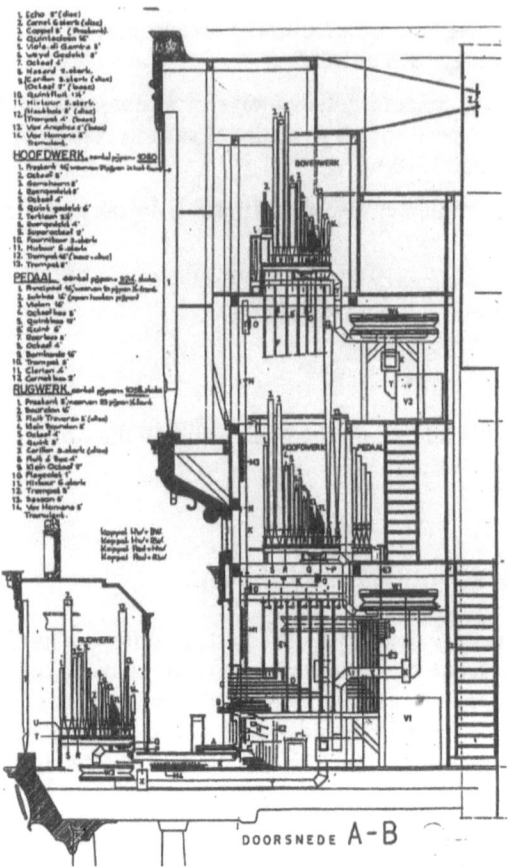

Figure 13.1 Characteristic internal stop disposition of Dutch organs, illustrated by the organ of St Stevenskerk, Nijmegen.

representative variety of organ tones from different cultures have been synthesised experimentally, namely:

- Flutes and Diapasons in the Dutch Classical tradition
- Diapasons in the modern Danish tradition
- Flutes in the German Baroque tradition
- Diapasons, Flutes and Reeds in the English Romantic tradition

Each of these applications places particular demands upon the synthesist. For example, the composition of Classical Mixture stops is much more complex than that of Romantic Mixtures, and requires the use of more synthesis resources. Certain varieties of Flute are very reliant for their effect upon a high level of characteristic wind-noise, both during the attack phase and in sustain. Within Dutch Classical organs, the relative disposition of different stops is very clearly defined (Figure 13.1), and is important to the production and perception of the organs' characteristic sound; this effect has to be accomplished within the synthesised waveforms. A variety of different temperaments have been used historically and regionally. Although the rich diversity of instruments makes for individuality rather than generality, some tonal trends can also be extrapolated: for example, German reeds tend generally to have a more developed partials' structure than their English counterparts, and French reeds are often scaled more heavily in lower registers.

For each of the applications, the experimental syntheses were assessed by organists and musicians in general, and by organ experts skilled in the traditions in question, to determine not only their acceptability as organ sound, but the degree to which the syntheses were able to represent the sounds of these varied traditions. For each tradition, the syntheses were judged to be successful on both counts.

13.6 Voicing Flexibility for Varied Acoustic Environment

An interesting example of using technology to promote accessibility to the cultural history of the organ arose in a recent project to install a multi-specification instrument at the University of Oxford. Oxford boasts a historic ceremonial concert hall, the Sheldonian Theatre, where the restoration or replacement costs for the existing pipe organ were judged to be, in the short term, prohibitive, so the decision was made to install a pipeless organ. Instead of just putting in a "pipe substitute", the opportunity was seized to provide something which pipes could not – a range of organ tone illustrating different historical styles and traditions, for use by organ scholars and in the wide variety of musical concerts performed at the theatre as well as on ceremonial occasions. BEST was chosen to fulfil the brief, because of its flexible voicing capability and the quality and complexity of its synthesis. The project was an ideal opportunity to test the capability of the system in bringing together the fruits of different organ cultures under one roof.

The specifications selected for inclusion in the project by the consultant, Simon Preston, were to illustrate the following styles:

- *English Romantic "Cathedral" style*: to be modelled on the organ built by Willis in 1876 at Salisbury Cathedral. At its installation this instrument was judged to be the finest organ in England. Willis was an innovative and idiosyncratic builder; the Willis sound includes high-pressure brilliant reeds and a keen-toned principal chorus which reduced in scale with increasing pitch. His voicings were in this respect very different from, for instance, Hill or Harrison, who also built large organs in the English tradition.

- *English "Classical" style*: to be modelled on the historic organ originally built by "Father" Smith in 1708 and restored/reconstructed by Mander at Pembroke College in Cambridge. The synthesis is tuned in English Ord temperament, as this has been identified as close to that likely to have been used by Smith.

- *French Romantic "Cathedral" style*: to be modelled on the organ built by the Aristide Cavaille-Coll in 1859 at Basilique Ste Clotilde in Paris. This instrument, for which Cesar Franck composed, retains all the original Cavaille-Coll voicing, despite later additions, and includes less strident pedal reeds than are often found in this builder's work.

A fourth specification in German Classical style is to be added later.

Upon examination of the chosen model instruments and comparison of their surroundings with those of the Sheldonian Theatre (see Figures 13.2–13.5), it is immediately apparent that to recreate the perceptual effect of these varied instruments is not just a question of recreating the waveforms which emanate from each pipe; the varied acoustics of the buildings in which the models are set has a massive influence upon the perception of the sound (Comerford and Comerford, 1995). The challenge is to use the flexibility of the technology to interpret the sound of the original through a different medium and in a different acoustic. In practice, some of the most critical waveform factors in successfully transferring sounds from one environment to another are the attack and decay transients, which can be accentuated by a more reverberant acoustic; the relative dynamics of the notes in each stop, and of the stops in each chorus, which may be amplified differentially by the building resonances; and the detailed partials' structure of specific stops, the perception of which can be altered by the presence or absence of, for example, close reflective surfaces. By virtue of using a synthesis system with the capability of adjusting all the

Figure 13.2 Salisbury Cathedral. Set within a massive building, the different departments of the organ are laid out on different sides of the choir, as shown in the photograph (and some in a transept), at the front of the Cathedral, at different levels, and with some pipes facing in different directions.

Figure 13.3 Pembroke College, Cambridge. The organ is in the centre of a flat gallery at the back of the small chapel; note the relative closeness of the reflective smooth plaster wall surfaces.

Figure 13.4 Basilique Ste Clotilde, Paris. The organ is constructed on a sloping gallery (the lowest part of which is about 60" above floor level), extending across the back of the tall but narrow Nave, with spacious chambers for each department. The right photograph is the view from the organ gallery.

Figure 13.5 The Sheldonian Theatre, Oxford. The magnificent Wren building is, internally, constructed entirely of wood and plaster (the apparent marble is a *trompe l'oeil*). The ceiling is canvas. Additional seating covers the wooden floor for concerts. The loudspeakers for the synthesis instrument have to be housed within the pipe organ case, as no visible alteration to the building is permitted.

components of each sound, these necessary alterations could be accommodated as required on each specification.

In this case, the syntheses were judged by the consultant to be a success, both to listen to and to play, in creating the impression of the sound of the original instruments within the new acoustic setting.

13.6.1 Player Interface

The use of a dedicated instrument as opposed to a midi keyboard solves the problem of providing a player-friendly and familiar user interface for the organist; but, if the flexibility of the synthesis instrument is to produce a *variety* of different organ specifications, then the organ console needs to be flexible too. Therefore for the Sheldonian Theatre instrument, a unique and fascinating console was developed by the installers, J. Wood and Sons of Bradford. The console has wooden disc push switches for stop selection, each with a central lamp to show when it is on, and a dedicated miniature display panel above each disc which shows the function of that switch on the current specification. Thus a switch which controls "Bourdon 16'" on one specification might act as "Viole de Gambe 8'" on another, the script on the display changing with the specification. The usage of the manuals is individual to each specification – on the English "Cathedral" instrument the Great department is the second manual, whereas on the French "Romantic" instrument the equivalent Grande department is played from the lowest manual. By such means the major features common to all organ traditions can be provided within a console which nevertheless has the flexibility to accommodate specialist features of different traditions. The console has a midi output facility.

13.7 Conclusions

At present, the use of a specialist synthesis technology with the flexibility to recreate sounds from different organ traditions is an effective way of using technology to widen access to the cultural history of the organ. It has been shown to successfully address the questions of sound quality, ensemble and voiceability in relation both to cultural diversity and acoustic environment, in a way of use to the student of organ sound. It also offers a solution to the player interface question. In the future, increasing Internet speeds and reducing costs should enable this synthesis solution to be playable from any location.

Acknowledgements

The authors are grateful to the European Commission for funding CRAFT project BE-S2-5285, which included some of the work reported in this paper, and to CRAFT colleagues for their collaboration.

References

Baker, D. (1991) *The Organ*. Shire Publications, Aylesbury.
Comerford, L. K. and Comerford, P. J. (1995) "The best stop on the organ" – Environmental acoustic and the perception of organ ensemble. *Proceedings of the Third International Conference on Acoustics and Musical Research*, Ferrara, Italy, May, pp. 375–380.
Comerford, L. K. and Comerford, P. J. (1996) Towards a design for pipeless organ loudspeakers. 12th Annual Conference in Reproduced Sound, Windermere, October. *Institute of Acoustics Journal*, 18(8), 219–226.
Sumner, W. L. (1981) *The Organ: its Evolution, Principles of Construction and Use*, 4th edn. Macdonald, London.
Wills, A. (1984) *Organ*. Macdonald, London.

14

Synthetic Vision for Road Traffic Simulation in a Virtual Environment

Wen Tang and Tao Ruan Wan

Abstract

In this chapter, we describe a method of using synthetic vision for road traffic simulation in a 3D virtual environment. We have implemented the synthetic vision techniques to cater for the dynamical features of the road traffic scenario by presenting an algorithm that is incorporated in the system to estimate the motion distributions of the scene at each animation frame. Multiple visual fields have been used to adjust and evaluate the driving conditions. Autonomous virtual vehicles in our system can make decisions based on a perceptual approach.

14.1 Introduction

3D virtual environments are very useful for simulating the behaviour of road traffic flow, which could provide rich information and quick solutions for traffic assessment. For example, a simulation can be used to assess road design, traffic monitoring and evacuation in an emergency. In the video game industry, driving and racing games are some of the most popular titles. Intelligent vehicles in a gaming environment can greatly enhance the realism and complexity of the gameplay. In reality, the behaviour of a road traffic flow is a complex phenomenon. There are a number of factors that can affect the behaviour of the flow:

- Volume of traffic on a road
- Highway codes
- Traffic control signals
- Speed of nearby vehicles
- Visibility of the road
- The behaviour of the drivers

Research has shown that the human visual perception plays an important role in human behaviour. The eyes actively perform problem-oriented selections and process the information from the visible world under the control of visual attention (Rybak *et al.*, 1998). The system described in this chapter is based on using synthetic vision techniques to provide visual information for behavioural modelling and control of the autonomous animated cars in a 3D virtual environment. The vision perceptual mechanisms of the autonomous cars in the system are the first communication layer between the internal abstract state of each car and its virtual environment in order to achieve behaviour animation. The motion behaviours of the autonomous cars and the flow of traffic are the end results of the perceptual and cognitive processes.

As one type of flock-like motion, road traffic flows share a similarity in behaviour with other types of flock, which can be found in humans or animals; however, they also have distinct characteristics, perceived as follows:

- *Collective behaviour*: a flow of cars on a road has so-called collective behaviour, formed by strong individualistic behaviours, which have tendencies to follow the flow and obey the traffic rules. Changing lanes and overtaking at certain desired speeds are examples of these internal mass behaviours of a car in a flow.

- *Multiple vision fields*: these are used effectively for heading judgements and decision making.

- *Manoeuvres and decision-making*: driving actions, such as entering a roundabout, overtaking and changing lanes, require full attention to the multiple visual information fields and will influence the behaviour of the traffic flow.

Based on these observations, we have developed an animation system to simulate the flow of road traffic by modelling the individual motion behaviours of a group of autonomous cars using a set of object-orientated simulation components. The vision component of the system is implemented as a set of moving visual fields with the ability to change the view volumes according to the driving conditions and specific driving tasks set by the system. The implementation takes advantage of computer graphics rendering techniques to improve the algorithm efficiency. Mirrors implemented as multiple view frustums are visual information resources for the driver's back and side views. In order to evaluate the dynamic changes in the traffic scene at each animation frame, information extracted from the visual fields has been analysed to direct the cars in the environment to behave autonomously. Therefore we are able to simulate the individual characteristics of the autonomous animated vehicles and the traffic flow behaviour effectively and simultaneously.

Section 14.2 reviews related work. Our vision system is described in Section 14.3. Section 14.4 describes the behaviour control and modelling mechanisms of the system. Animation results are shown in Section 14.5, and Section 14.6 discusses the work done so far and the future directions for this research project.

14.2 Related Work

Influenced by the research work done in imaging and vision computing in the robotics area, a number of behaviour animation systems have been developed that have had one or more kinds of visual perception models implemented.

Reynolds has designed a simple yet effective perception model and used it in his pioneer behaviour animation system – the "flock of birds" simulation (Reynolds, 1987). A spherical zone at the local origin of each bird in the simulation is a sensor for detecting collisions from its nearby flockmates. Simulated birds have direct access to the geometric database to get the position, orientation and velocity of all other objects in the environment. A similar approach has been adapted by Tu *et al.* in their system for animating a group of artificial fish. The fish have a limited field of view (Tu and Terzopoulos, 1994). These sensor-based approaches are effective in generating localised behaviour information; however, the efficiency of the algorithm suffers from the increasing number of objects in the virtual world. In general, an individual agent (e.g. bird or fish) has to process every other object in the environment to avoid collisions and to query information for animation; the complexity of the flocking algorithm is basically $O(n^2)$. This algorithm can become too inefficient for a computer game which demands real-time interactions or for use in a real-time simulation system with a large complex virtual environment. Renault *et al.* (1990) developed a synthetic vision technique that receives information from the view frustum of a virtual human. In their system, a virtual human has a single visual field for avoiding environmental obstacles. The use of synthetic vision, in some sense, is closer to the way in which humans perceive information from the environment than that of the sensor-based approach, since humans do not always possess all the information about their environment. The efficiency of the synthetic vision method can be improved by taking advantage of computer graphics system hardware and graphics rendering algorithms. Blumberg (1996, 1997) has implemented a motion energy algorithm to analysis the images of a virtual dog's single visual field to direct the dog to avoid environmental obstacles in a virtual world.

Compared with the behaviour animation systems reviewed above, a virtual traffic simulation is different in that autonomous cars have multiple visual fields and the motion behaviours of the cars are the end results of continuous attention to these fields. Collision avoidance of the environmental obstacles can be done effectively by simply programming the cars to "follow the road" layout vectors. One of the most important issues of a traffic simulation is how to adjust the road situation for motion manoeuvres.

14.3 The Vision System

Simulations of the autonomous behaviours of cars in a 3D virtual environment are only approximations of real-life situations to a certain extent. At this stage, in order to decompose the complexity of the scene, assumptions have been made about weather conditions and visibility in the environment, and a set of highway codes has been adapted.

The details of our vision system are described in the following subsections.

14.3.1 Multiple Visual Fields

Each vehicle in the system is assigned four visual fields, as shown in Figure 14.1. A moving camera is attached to a car for its front view. The angle of the front field of view can be changed dynamically according to the traffic situation. Approaching a roundabout requires a larger angle than that needed for driving on a straight road. Each car has three "mirrors" attached to it for the back and side views respectively. A user-controlled car has been used in our system to input external stimulus into the virtual environment and increase the complexity of the behaviour simulation.

At each animation frame, for each car, visibility tests are carried out to detect which cars in the scene fall into its visual fields. This test can be effectively done by attaching bounding spheres to the cars in the scene. For every other car in the scene, we test its bounding sphere against each view frustum's five planes of the examined car. If the bounding sphere of a car is in front of any of the planes of a view frustum, the car is in that view frustum of the examined car. Hence it is visible to the examined car.

As shown in Figure 14.2, cars 1, 2 and 3 are completely inside the back view frustum of the examined car, while car 4 is only partially inside the frustum. Car 6 is completely outside the frustum. Therefore in this example, cars 1, 2, 3 and 4 are visible to the examined car. However, testing the visibility of each car in the scene with four view frustums of every other car in the scene can be a costly operation. As the number of cars in the virtual environment increases, the rendering frame rate will plummet rapidly due to the processing expense of

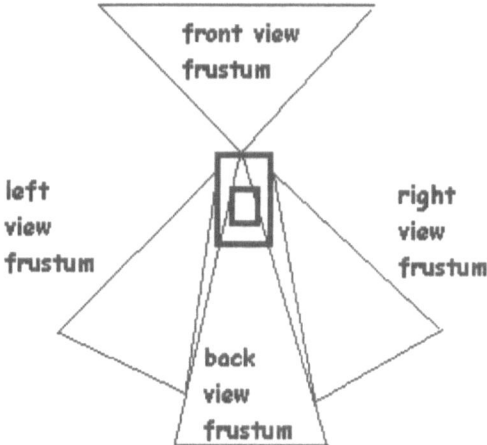

Figure 14.1 Multiple fields of view.

test car back view frustum

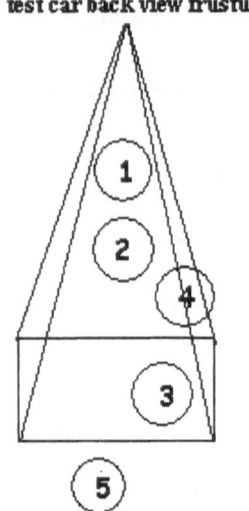

Figure 14.2 Detecting the bounding spheres with the test car's back view frustum.

the visibility tests. In this case, the test algorithm is $4 \times O(n^2)$. In order to increase the efficiency of the method, in our system two nested bounding spheres are attached to each car (Figure 14.3).

The outer bounding sphere includes the car and its four view frustums, whilst the inner bounding sphere includes only the car itself. We test the inner sphere of every other car with the outer sphere of the examined car to decide whether

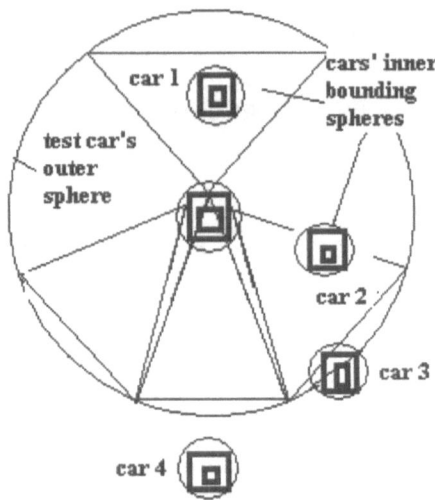

Figure 14.3 Detecting the examined car's outer bounding sphere with the others' inner bounding spheres.

or not there is a need for further visibility tests with the view frustums of the examined car. If the two spheres collide or the inner sphere is inside the examined car's outer sphere then there is need for further tests; otherwise no visibility test is required: the car is invisible to the examined car.

As shown in Figure 14.3, cars 1, 2 and 3 require further visibility tests, whilst car 4 is outside the outer bounding sphere of the examined car. Hence there is no need for further tests for car 4.

14.3.2 Evaluating the Visual Field Information

The internal state structure of each autonomous car in the system has been designed as a *finite state machine* (FSM) in which the car's state changes from state to state based on its current state and the input it currently receives. At each animation frame, the state of the car has a specific set of outputs.

The internal state of each car is a behaviour structure which contains information about the car such as its speed, direction, acceleration, driving path, current driving direction and the current state of the car. The state is updated constantly during the animation. Hence, at a given animation frame, for each car that has fallen into the visual fields of the examined car, the state information of that car can be obtained. The values obtained from each of the view frustums of the examined car are calculated for comparison with a number of thresholds for decision making. The driving attenuation coefficients in each visual field are calculated to assess the overall driving conditions in that direction. The greater the value, the lower are the velocities and accelerations. The motion influence of each car's mass behaviour to other cars in the flow is in approximately inverse proportion to the square of the distances between the cars. We can simply discard the operations of changing the motion state for any car whose distance from the examined car exceeds a distance threshold for the purpose of algorithm efficiency. Geometric information about the object reflected in the "mirrors" of the cars can be obtained by a reflection transformation matrix to the "mirrors'" local origins, which are attached to the local system of the autonomous car. We can directly access the internal states of the cars that fallen into the view frustums of the three mirrors in the same way as we did with the front view frustum. The velocities obtained from each of the fields are normalised. The results of visual information analysis are then used to control the underlying locomotion of the cars.

In general, a driver mostly pays attention to the changes of driving signals from the cars that are directly next to his or her car in all directions. The manoeuvring information of the nearby cars can be obtained from the object database, which is captured by each visual field over a sequence of frames.

14.4 Behaviour Model and Motion Control

As described above, the internal state of each car in the scene is designed as an FSM in which a set of rules have been defined to implement the self-contained

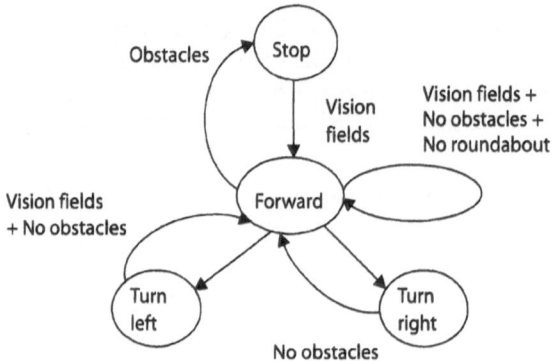

Figure 14.4 The FSM state of the autonomous car.

autonomous properties of the car. Figure 14.4 illustrates the cars' FSM mechanism. The transition between the states of a car is performed by the behaviour modelling functions. The flow behaviour of a traffic scene is the result of the individual behaviour of cars on the road. A behaviour modelling system has been designed to update the internal state of the autonomous animated car consistently. The modelling system is a component of a group of functions that query the vision system at each animation frame and evaluate the visual fields of the autonomous car. A motion filter analyses each visual field and outputs information related to the speed forces and steering forces to accelerate/decelerate and change the direction of the autonomous cars. The behaviour modelling system also modifies the global state of the car, as shown in Figure 14.5.

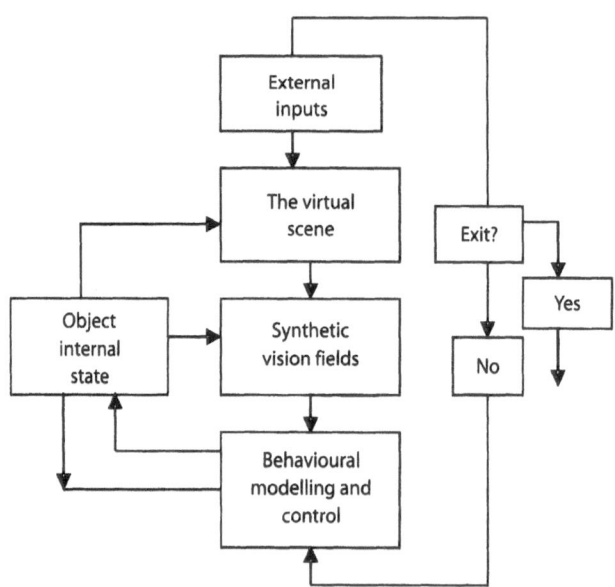

Figure 14.5 Behaviour control and modelling process.

14.5 Results and Discussion

We have designed a simulation scenario which contains a group of cars and some simple road layouts. A number of driving actions have been tested, including driving on a straight road and turning at a roundabout.

Figure 14.6 is a screenshot of a traffic simulation system to illustrate autonomous cars driving on a straight road. Figure 14.7 shows the cars approaching/ entering a roundabout. Figure 14.8 demonstrates the simulation scenario with three small built-in rendering windows to represent the GREE car's back, left-side and right-side visual fields respectively. Figure 14.9 shows a set of sequences of changing visual information captured by the back, left-side, and the right-side view frustums of the examined cars.

Figure 14.6 Autonomous animated cars driving on a straight road.

Figure 14.7 Autonomous animated cars approaching a roundabout.

Figure 14.8 A simulation scene with three built-in rendering windows for the examined car's back, left-side, and right-side visual fields.

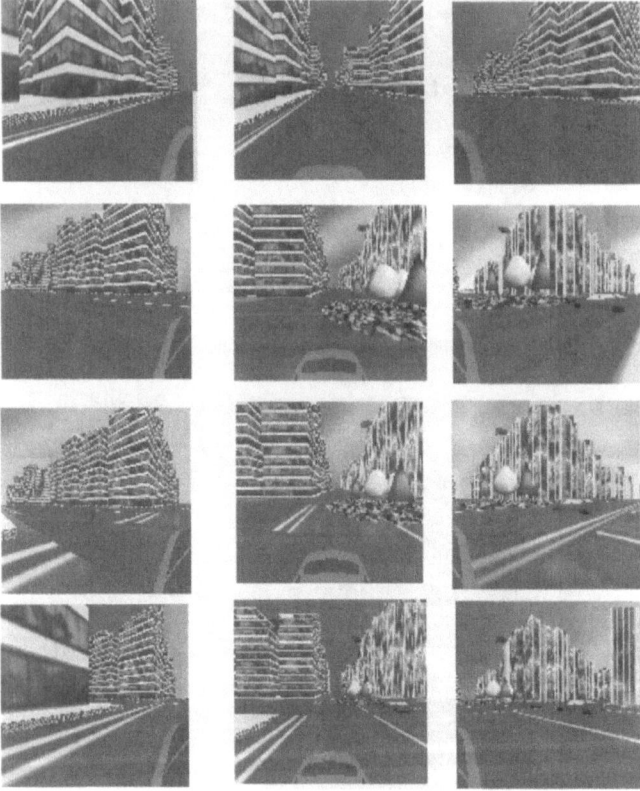

Figure 14.9 A set of sequences of the changing information in the back, left-side, and right-side visual fields of the examined car at different animation frames.

14.6 Conclusions and Further Work

We have presented a traffic simulation system using synthetic vision techniques to obtain road information for the control of autonomous cars in a virtual environment. The methods discussed here are effective and the simulation results are realistic. Currently the behavioural modelling system described here is a primary system that needs to be extended to model distributed vehicle behaviours. The vision system also needs to be improved for processing more complex and general traffic scenes for real-time simulations. A further research direction is to improve the system so that it could include the activities and behaviours of pedestrians on the roads in it. Such a system will be a more sophisticated and much closer simulation of the real world.

References

Blumberg, B. (1996) Old tricks, new dogs: ethology and interactive creatures. *Ph.D. Thesis*, The Media Laboratory, Massachusetts Institute of Technology, Cambridge, MA, September.

Blumberg, B. (1997) Go with the flow: synthetic vision for autonomous animated creatures. *Poster*, AAAI Conference on Autonomous Agents, Marina Del Ray, CA.

Renault, O., Magnenat-Thalmann, N. and Thalmann, D. (1990) A vision-based approach to behavioural animation. *The Journal of Visualisation and Computer Animation*, 1(1), 18–21.

Reynolds, C. W. (1987) Flocks, herds, and schools: a distributed behavioral model. *Proceedings of SIGGRAPH '87*, Anaheim, CA, 27–31, July. In *Computer Graphics*, 21(4), 25–34.

Rybak, I. A., Gusakova, V. I., Golovan, A. V., Podladchikova, L. N. and Shevtsova, N. A. (1998) A model of attention-guided visual perception and recognition. *Vision Research*, **38**, 2387–2400.

Tu, X. and Terzopoulos, D. (1994) Artificial fishes: physics, locomotion, perception, behavior. *Proceedings of SIGGRAPH '94*. Computer Graphics Proceedings, Annual Conference Series, pp. 43–50.

15
Eternity

Jon Pettigrew

Abstract

Eternity is a "Gedanken" digital arts project for children. It explores the usage and attitudes of children using computers for creative collaboration. Computers are providing a new way for children to be creative and new generations of artists will have different expectations.

Generative software systems are emancipatory and this changes the way in which adults and children interact, as well as the human–computer interaction issues. Children actively seek collaboration and explore intelligent user interfaces with ease. Computers help us to think creatively (Edmonds, 2000).

Barriers to the use of creative digital arts systems are more likely to be human and administrative than artistic. Children in the study were not open to Internet collaboration. The Internet is more of a database technology than a social technology.

Providing chances for children to use computers for creative purposes can alter the perception that computers are simply efficiency tools, training people for the new factory work of data input and telephone call centre workers.

15.1 Introduction

Children are different from adults. As collaborators with computers they come with open minds and are much more likely to explore and have the confidence to experiment. These important principles, outlined in Pettigrew and Elliott (1999), are one strand of the research into children's computer creativity.

A short review of books about the use of computers in education and with children published across a relatively short space of time shows the changing nature of Human-Computer Interaction (HCI). White and Hubbard (1988) predicted that "teachers will spend most of their time guiding and counselling and leading small teams rather than lecturing to a room full of students". They continue: "It is clear that the computer is the most powerful tool ever developed and that children need to learn how to use them in a variety of ways. It is likely that computers will change the way in which we learn and part of this is increasing learner control and choice".

In *Developing CAL: Computers in the Curriculum*, Watson (1997) uses a model which can help to explain the context of computers for creative purposes and some of the barriers to their successful use.

He outlines four basic modes of computer use in Computer-Assisted Learning:

- Instructional
- Revelatory
- Conjectural
- Emancipatory

Each of these modes has its uses, its product types and its place in the history of CAL. The initial use of computers, and one that is still widely practised, is of Instructional "skill and drill" programmes, with limited choices and based on the Behaviourist model of learning. Games and simulation are included in the Revelatory mode; databases and the World Wide Web are the key types in Conjectural modes. We have now reached the level of wider availability of Emancipatory systems for creating new content. Only with more memory and speed can the emancipatory mode be widely used. But hardware developments are not the sole predictors of computer usage and attitudes.

What Watson contends is that the move from subject and content concentration towards learning-centred activities, where there is little or no content, is c.i ical. This is a move from the "computer programming the child" to "children programming the computer". The *Eternity* project is investigating, with one small project, what differences there are in children being in control rather than the computer being in control.

Some of the barriers to successful use of ICTs may be highlighted by these kinds of investigation. *Eternity* is a useful paradigm to discover what influences the New Media Technologies (NMTs) have on creativity in art-practice with children and to identify what requirements such systems should have in the future.

It should not be forgotten that computers are not the answer to all our educational prayers. Even Seymour Papert (1982, 1999) would not suggest that simply introducing computers into children's art practice in classrooms would necessarily increase creativity. Papert, when a senior early figure in MIT's Media Lab, was lobbying for widespread introduction of computers for children in school. He anticipated that this would enable them to understand computers and programming via his Logo language.

It is interesting to note now that the leading neuroscientist Susan Greenfield (quoted in Miller (2000)) at a Royal Institution Address warned that the standard information received from computers could lead to "standardised brain connections" and "a stifling of individuality and creativity".

Children's art has long been an area of investigation and wonder. Children are motivated by creating new things, rather than simply consuming other people's work. In the visual arts, children's art can be a source for modern artists and it can be a starting point for art teaching; for some it is a metaphor and for others a lever into the human mind. Jonathan Fineberg (1998) in his introduction to

Discovering Child Art writes children's art is "something we all have in common, and its pervasive influence on modern (ie adult) art speaks to the fundamentally human qualities that motivate the work of modern artists".

Art is encoded thought. Music is distributed emotion. Children, like adults, respond to music. Music has become a pervasive influence and is no longer a special part of people's lives, it surrounds it. The 20th century has been referred to as the Ambient Century for this reason. People like to appreciate music, but they also like the idea that they can be creators too.

Art and technology have developed closer links. ICTs are no longer simply seen as efficiency tools: they can also be used to help people express themselves. The role of computers as catalysts or stimulants to human thinking and content generation, suggest that computers can help us think creatively (Edmonds, 2000). Art and creativity supported by computers is an area of study within the human–machine interface which should become increasingly important. Understanding children's approaches and values in this context is a valuable part of HCI.

15.2 Method

The *Eternity* project has started to consider the role of children as musical artists supported by smart software in the Emancipatory mode. It is a "Gedanken" art project, that is an art "thought-experiment". Those investigating Artificial Intelligence (AI) use Gedanken experiments, such as the "Chinese Room" (Searle, 1980), and they were a method pioneered by 19th century German psychologists. Children are intrigued by thought experiments, and they are equally intrigued by concepts of large numbers of objects and limitlessness.

There are four stages to the *Eternity* process, leading to children aged 10–12 trying to answer the question "How can we create a piece of music to last for ever?".

Firstly, children are introduced to the software programme SSEYO Koan X, which is constructed to enable the mixing of musical fragments or templates into appealing pieces which play generatively.

Koan X is a computer programme which has built in smarts for many musical features, and by exploration and experimentation, children can develop musical compositions without the need for prior knowledge or detailed training. The "musical intelligence" provides opportunities for children to show aesthetic judgement and develop creativity in themselves and with others. They can make music either for their own purposes or to exhibit to audiences.

The "X" factor is a device which enables the mix to change generatively, as well as the fragments and templates being generative themselves. This means that the pieces alter subtly each time they are played and to all intents and purposes they are never the same twice. Cockerton *et al.* (1997) used Koan in a series of experiments on performance and background music, suggesting that its "non-repetitive" nature helped improve performance in cognitive tasks.

Eno (1997) first described the concept of "generative" applied to art in the context of Koan. Generative art is described on the http://www.generative.net/ site as "a term given to work which stems from concentrating on the processes involved in producing an artwork. Usually (although not strictly) automated by the use of a machine or computer, or by using mathematical or pragmatic instructions to define rules by which such artworks are executed". The debate continues via the "eu-gene" mailing list.

Secondly, children discuss concepts of Infinity and Eternity. Thirdly, they start to create a piece of music that could last forever. Finally they set up the piece to play for Eternity.

The first stage is normally relatively short, but can be extended to include revisions and playing pieces to an audience. In other words this stage can be part of a general musical education, getting to know the way in which the software works and the way in which the changes to instruments (patch), scale, key, volume, pan and speed can have different effects on the pieces.

The second stage needs to be well managed. There are in fact a number of different concepts of "largeness" each associated with a different mathematical symbol. Using a Latin description there is "in terra", which describes a large number of things (∞), "in abstracto" for mathematical sets (ω and aleph zero) and "in Deo" for time (Ω).

Children enjoy responding to questions which tax their minds because there are no real solutions to these problems. *Koan* is in fact a Japanese word for a Zen Buddhist concept or series of concepts, the most familiar of which are questions such as "What is the sound of one hand clapping?" which create meditative frames of mind. These are referred to as Koans. "The student is to actively and intensively inquire, search, and look into the koan, raising in himself a fiery and attentive attitude known technically as the I-Ching" Wilber (1993).

A grid of different types of questions was created for use with the children. They added new questions to the core set which can then be passed on to other children. Examples follow in the Section 15.3.

The two participating communities are Junior Schools in Hove in Sussex and Trimdon, County Durham, UK. The initial phase in Hove was as part of an after-school club with four main 10-year-old activists, which was then transferred to a Year 6 (10 and 11 year olds) class activity during an Arts Week. A full school Assembly was conducted to explain the project and for the school to listen to the music the children had produced. Shortly before the start of the Hove project, a group of children at the Trimdon 2000 summer school worked on the same idea. The children sent their pieces to an email account which had a shared password, so that children could access the pieces and, from an open file format, alter the pieces from the other children and the corresponding projects.

Each community had other chances to work on the *Eternity* project and both were told of the opportunity for an Arts Centre in New York to become involved.

From August 1999 to September 2000, about 30 children participated in the project and in late September 2000, four children from the Trimdon project visited the school in Hove.

Trimdon 2000 is part of the Trimdon Digital Village developments. The village has become networked at various locations, including the schools, community centre, library, pubs and Labour Club, enabling a wide variety of digital information and communications activities to take place. The infrastructure and project planning has been supported by the Community Informatics Research and Applications Unit (CIRA) of Teesside University. One of their technicians was a key helper on the Technical Support side of *Eternity*. Hove is part of the new City of Brighton and Hove, which is a leading centre for new media technologies and content production as well as the main centre of "Wired Sussex".

15.3 Results

Children can achieve using emancipatory smart software. The *Eternity* children were all able to quickly use the standard GUI to understand what the software could do. They used a primitive intelligent user interface with ease. Appealing and satisfying musical compositions were made quickly and widened the appreciation of what computers can do.

There is a gap between many adults and children in their feelings towards computers and what they are capable of. Children view them much more like appliances, not special machines that you have to be trained on. One teacher characterised this by insisting that if she got involved she would need a day's training in the use of the software, whereas most children can use Koan X in two minutes. Other practical barriers were difficulties in configuring the machines to use and play music via soundcards, avoiding other children "misusing" the setups afterwards and between sessions. Open access to email for children is still difficult for organisations to set up and monitor.

Increasingly, there is no longer a fixed body of knowledge or years of good practice orthodoxy which have to be instilled into learners from "knowers"; there is a framework and projects are created. In the context of *Eternity*, children were starting to be creative and collaborate with each other and Koan in diverse and interesting ways.

Children are artists (Figure 15.1), but this activity is just one of very many that they do; unlike adult artists children do not remain on task for long. Computer-mediated creativity is just one part of their rich tapestry of life. Their responses seem to suggest that children expect and should get many chances to explore digital creativity, but they will not all become artists in the traditional sense of the word.

Children find it a lot easier to collaborate with each other in their own communities and friendship groups than via the Internet. Even when the children from the two UK projects met, there were many human processes they needed to go through before they could consider the artistic collaboration issues. The idea of

Figure 15.1

working with children in New York was very appealing, but they were more interested to know answers to questions about where they lived and what they ate than going straight into music collaboration. Unfortunately, the musical stage was not reached, and the "evangelist" in New York had too many practical problems to deal with as well.

The connections between the three communities did not develop apart from an initial interest in hearing the pieces. Many children at this age appear to find it difficult to develop positive Internet relationships and understand the mechanics and reasons for doing so.

To set up the *Eternity* project, the second stage of asking questions provided some fascinating inspiration for the children. These are some of the questions and answers that were created:

- How many people do you know?
- How many people are there in the world now?
- How many different sorts of animal are there in the world now?
- How many places have you visited?
- How many people live to be 100 years old now?
- How many people will live to be 100 years old in 100 years' time?
- How long will you live for?
- What can last forever?

The list is endless.

Q How far is it to the end of the Universe?
A You don't ever get to the end.

Q How long is eternity?
A I don't know, till the end of the Earth.

Q Why don't you want to live forever?

A I'd like to live forever but stay young.

Q How could we make a piece of music last forever?
  - A robot orchestra.
  - Tape it.
  - Press repeat on a CD player.
  - Make sure you have lots of money and fill the electricity up.
  - Write it out, just in case.
  - You could email people the music.

Q How would you use Koan to make a piece of music last forever?
  - Press space bar.
  - A very long drum beat.
  - If there weren't computers you couldn't have a Koan which lasts for ever.
  - Who will switch the computer off?
  - Put it on a radio station and call it Eternal Music FM.

Q How many computers would you need?
  - Two, so when you wake up and notice one is working you can buy another one. Yes, like a mum and a dad computer.

This process allowed children to have a reason to use Koan X collaboratively to try to produce a piece of music to last forever. It provided a suitable frame in which they can be creative and working together to develop, alter and listen to pieces of music which reached the project aim and their own interests.

Listening to their pieces, fragments and trials is the best way to consider their success, pieces such as:

- Drumalot.skd
- Night-time bumper.skd
- The peaceful sea.skd
- Getting jiggy with it.skd

.skd stands for sseyo koan design, the file format which stores the data as a set of instructions which can then be sent as a small file; an infinity of music in 12 kbyte of code.

Figure 15.2

Computer-based creativity allows more children to participate: it is easier for them and enables them to develop aesthetic judgement about what is good and for what purpose. Children can act more independently and be less reliant on formal teaching methods.

Eternity suggests that children can develop repertories of behaviour which can be transferred to other creative purposes (Figure 15.2). For instance, one group of children later used the Koan system to write musical soundscapes for a staging of *The Tempest*. Collaboration and discussion between children with the computer is a natural part of the art-practice for them and not just because there are rarely enough computers for children to work on their own; they seek collaboration.

15.4 Discussion

In some ways the project so far is a pilot investigation into the role of computers in creativity in children, and as such it suggesting a number of outline ideas which are being further researched. The approaches taken by Candy (1999) and Scanlon *et al.* (1999) are helping to define the Action Research approaches to systematically investigate the nature of the HCI between children and computers for creative art-practice.

The information age is not just about using ICTs more; it is about changes in the nature of knowledge and processes which fundamentally mean new generations of computer users with different expectations. The information age will not be more of what we know; it will be a change from fixed content and process approaches to generative systems where teachers will not be the custodians of the process. Exploration and experiment, as well as dealing with the feelings about ICTs will "need to be taken into account" (Pettigrew and Elliott, 1999). Fixed outcomes, like the UK's National Curriculum will need to be adapted by teachers to deal with these changes. There is a need for digital art-practice evangelists and people with confidence and openness to new ideas and projects.

The human issues are still the major constraints for this project. In each of the communities, once the children are set and centred on the task the creativity starts, and while they are centred and motivated it lasts. Practical and technical considerations need to be well managed, otherwise the opportunities do not come about at all. The evangelists need to be aware of the hardware, software and networking ecologies of the places they work in to make the creative experience as effortless as possible. Children themselves are adaptable and want to learn and experience what the New Media Technologies offer them, but even if their communities are supportive there can be practical issues which stop the discussion of the project developing into unplanned ways.

The Internet has passed the early adopter stage, when users were pleased to try anything out. As a natural part of ICT and Entertainment now, getting used to the rules of distant interaction of the Internet is an important part of life-skilling. Children are warned against the dangers of the Internet; therefore they need support and proof of interest before they can use the Internet for collaborative arts purposes. The move from information storehouse to social technology is slow – "the software for human interaction are less developed then database technologies" Wallace (1999) – suggesting that the techno-push mode needs to consider what children really think and do, at least for the time being.

As computers have moved from the innovators and early adopters amongst teachers to mainstream National Curriculum requirements (in the UK), the evangelists of Papert's Logo computer programming tool for children are no longer the drivers of policy and applications of IT for children in the classroom and the community. Children with intelligent user interfaces now do program computers, but do not code them. Nevertheless there are still areas worth exploring and fighting for and creativity is one of these. Otherwise computers will simply be characterised as efficiency systems to manipulate data in fixed patterns for spreadsheet, database and text manipulation. Miller (2000) writing in *The Guardian* sees this approach as providing training only for "the increasing numbers of people in white collar, low skill jobs as data input clerks and telephone centre workers".

If a tree falls in the distant forest does it make a sound?

References

Candy, L. (1999) COSTART Project Artists Survey: Preliminary Results. *LUTCHI:C&CRS Research Report*, November, Loughborough University

Cockerton, T. *et al.* (1999) Cognitive test performance and background music. *Perceptual and Motor Skills*, 85, 1435–1438.

Edmonds, E. A. (2000) Artists augmented by agents. In *International Conference on Intelligent User Interfaces* (ed. H. Lieberman). ACM Press, New York, pp. 68–73.

Eno, B. (1997) *A Year with Swollen Appendices*. Faber & Faber, London.

Fineberg, J. (1998) *Discovering Children's Art*. Princeton University Press, Princeton, NJ.

Miller, B. (2000) Mind games. *The Guardian*, 13 December, Section 2 pp. 10–11.

Papert, S. (1982) Tomorrow's classrooms. *Times Educational Supplement*, 5 March, pp. 31–32, 41; available at http://www.papert.org/articles/TomorrowsClassrooms.html.

Papert, S. (1999) Ghost in the machine. *ZineZone.com*; available at http://www.papert.org/articles/GhostInTheMachine.html

Pettigrew, M. and Elliott, L. (1999) *Student IT Skills*. Gower, London.

Searle, J. R. (1980) Minds, brains and programs. *The Behaviour and Brain Sciences*, 3, 417–457.

Scanlon, E. *et al.* (1999) Collaboration in a primary classroom: mediating science activities through new technology. In *Learning with Computers* (eds. K. Littleton and P. Light). Routledge, London.

Wallace, P. (1999) *The Psychology of the Internet*. Cambridge University Press, Cambridge.

Watson, L. (1997) *Developing CAL: Computers in the Classroom*. HarperCollins, London.

White, J. and Hubbard, D. (1988) *Computers and Education*. Macmillan, London.

Wilber, K. (1993) *The Spectrum of Consciousness*. First Quest Editions, Wheaton.

About the Author

Jon Pettigrew is a researcher with the Creativity and Cognition Research Studios at Loughborough University, supervised by Professor Ernest Edmonds. He gained his first degree at Newcastle University in Psychology and has a Masters from Cranfield

University in Business Administration. He co-founded SSEYO Ltd in 1990, the company
which developed the Koan software system, and left in 2000. Since then he has lectured
in psychology and continued with his consultancy work for a variety of clients. He is a
member of the British Screen Advisory Council.

Author Index